Engineers' Guide to Rotating Equipment

The Pocket Reference

Clifford Matthews
BSc, CEng, MBA, FIMechE

Professional Engineering Publishing Limited,
London and Bury St Edmunds, UK

First published 2002

ISBN 1 86058 344 X

A CIP catalogue record for this book is available from the British Library.

This book is intended to assist engineers and designers in understanding and fulfilling their obligations and responsibilities. All interpretation contained in this publication – concerning technical, regulatory, and design information and data, unless specifically otherwise identified – carries no authority. The information given here is not intended to be used for the design, manufacture, repair, inspection, or certification of pressure equipment, whether or not that equipment is subject to design codes and statutory requirements. Engineers and designers dealing with pressure equipment should not use the information in this book to demonstrate compliance with any code, standard, or regulatory requirement. While great care has been taken in the preparation of this publication, neither the author nor the Publishers do warrant, guarantee, or make any representation regarding the use of this publication in terms of correctness, accuracy, reliability, currentness, comprehensiveness, or otherwise. Neither the Publisher, Author, nor anyone, nor anybody who has been involved in the creation, production, or delivery of this product shall be liable for any direct, indirect, consequential, or incidental damages arising from its use.

Contents

About the Author

Cliff has extensive experience as consulting/inspection engineer on power/chemical plant projects worldwide: Europe, Asia, Middle East, USA, Central and South America, and Africa. He has been an expert witness in a wide variety of insurance investigations and technical disputes in power plants, ships, paper mills, and glass plants concerning values of $40 m. Cliff also performs factory inspections in all parts of the world including China, USA, Western and Eastern Europe. He carries out site engineering in the Caribbean – Jamaica, Bahamas, and the Cayman Islands.

Cliff is also the author of several books and training courses on pressure equipment-related related subjects.

PREFACE

How to Use this Book

This book is intended to be an introductory guide to rotating equipment, suitable for use as a 'first port of call' for information on the subject. It tries to incorporate both technical and administrative aspects of rotating equipment manufacture and use, introducing the basic principles of balancing, vibration, noise, and inspection and testing of a wide range of equipment. There is some well-established content and a few newer ideas. It makes references to the most commonly used current and recent pressure technical codes and standards, and attempts to simplify their complex content into a form that is easier to understand. By necessity, therefore, the content of this introductory book is not a substitute for the full text of statutory instruments, regulations, and technical codes/standards. In all cases, reference must be made to the latest edition of the relevant document to obtain full, up-to-date information. Similarly, technical guidelines and 'rules of thumb' given in the book should be taken as just that – their only purpose is to be useful.

This introductory guide to rotating equipment is divided into 14 main chapters covering practical, theoretical, and legislative aspects of rotating equipment technology. Content includes website and documentary references for technical and regulatory information about rotating equipment design and manufacture. Formal design-related information appears in the referenced sources, while the websites provide a wide spread of related information that can be used on a more informal basis. Most information that you will need can be obtained from the websites, if you know where to look.

Chapter 1 provides details of engineering units systems and mathematics, essential to understanding the principles on which rotating equipment performance is based. The basics of statics and deformable body mechanics are given in Chapter 2, leading on to Chapter 3, which covers motion and dynamics. The generic topics of balancing, vibration, and noise are introduced in Chapter 4; these are common to virtually all types of rotating equipment. Chapter 5 provides an outline of the various machine elements that make up rotating machinery. Chapter 6, covering fluid mechanics, is a necessarily theoretical chapter, providing formal explanations of essential fluid mechanics principles used in the design of rotating fluid machinery. Individual types of rotating equipment such as pumps, compressors, turbines, and their associated power transmission equipment are outlined in Chapters 7–10. Chapters 11 and 12 are practically orientated, looking at the basic principles of mechanical design and material choice used in the design of all types of rotating equipment. In common with other areas of mechanical engineering, there have been rapid legislative developments over the past few years; Chapter 13 provides detailed summaries of the content and implications of The European Machinery Directives, and mentions the proposed 'Amending Directive 95/16/EC' that may cause further changes in the future.

Finally, the purpose of this introductory book is to provide a useful pocket-size source of reference for engineers, technicians, and students with activities in the rotating equipment business. If there is basic *introductory* information about rotating equipment you need, I want you to be able to find it here. If you have any observations about omissions (or errors) your comments will be welcomed and used towards future editions of this book. Please submit them to:

Ukdatabook2000@aol.com

If you have any informal technical comments you can submit them through my website at: www.plant-inspection.org.uk

Clifford Matthews BSc, CEng, MBA, FIMechE

INTRODUCTION

The Role of Technical Standards

Technical standards play an important role in the design, manufacture, and testing of rotating equipment components and machinery. In many cases, rotating machines use a wide variety of types of technical standards: complex, theoretically based topics for kinetic and dynamic design complemented by more practical engineering-based standards for materials, manufacture, non-destructive and pressure testing. Published standards also have wide acceptance for vibration measurement and dynamic balancing of rotating components and systems.

In common with other areas of mechanical engineering, rotating equipment is increasingly subject to the regime of EU directives and their corresponding harmonized standards. In particular, The Machinery Directives are now well established, with wide-ranging influence on design, manufacture, operation, and maintenance documentation. Harmonization is not an instant process, however, and there are still many well-accepted national standards (European and American) that are used as sound (and proven) technical guidance.

Because of the complexity of rotating equipment, technical standards relating to basic mechanical design (mechanics, tolerances, limit and fits, surface finish, etc.) continue to be important. These standards form the foundation of mechanical engineering and are based on sound experience, gained over time.

One area of emerging technical standards is that of environmental compliance. Most types of engines and prime movers come under the classification of 'rotating equipment' and these machines are increasingly subject to legislative limits on emissions and noise. Health and Safety requirements are also growing, with new standards emerging covering machine safety, integrity, and vibration limits.

As in many engineering disciplines, technical standards relevant to rotating equipment use several systems of units. Although the Système International (SI) is favoured in Europe, the USA retains the use of the USCS 'imperial' system, as do many other parts of the world. There are also industry-specific preferences; the aerospace and offshore industries are still biased, in many areas, towards imperial units-based technical standards. These industries are big users of gas turbines, and other complex fluid equipment.

In using the information in this book, it is important to refer to the latest version of any published technical standard mentioned. New standards are being issued rapidly as the European standards harmonization programme progresses and there are often small and subtle changes in new versions of previously well-established technical standards.

CHAPTER 1

Engineering Fundamentals

1.1 The Greek alphabet

The Greek alphabet is used extensively in Europe and the United States to denote engineering quantities (see Table 1.1). Each letter can have various meanings, depending on the context in which it is used.

Table 1.1 The Greek alphabet

Name	*Symbol*	
	Capital	*Lower case*
alpha	A	α
beta	B	β
gamma	Γ	γ
delta	Δ	δ
epsilon	E	ε
zeta	Z	ζ
eta	H	η
theta	Θ	θ
iota	I	ι
kappa	K	κ
lambda	Λ	λ
mu	M	μ
nu	N	ν
xi	Ξ	ξ
omicron	O	o
pi	Π	π

Table 1.1 Cont.

rho	P	ρ
sigma	Σ	σ
tau	T	τ
upsilon	Y	υ
phi	Φ	ϕ
chi	X	χ
psi	Ψ	ψ
omega	Ω	ω

1.2 Units systems

In the United States, the most commonly used system of units in the rotating equipment industry is the United States Customary System (USCS). The 'MKS system' is a metric system still used in some European countries, but gradually being superseded by the expanded Système International (SI) system.

The USCS system

Countries outside the USA often refer to this as the 'inch–pound' system. The base units are:

Length: foot (ft) = 12 inches (in)
Force: pound force or thrust (lbf)
Time: second (s)
Temperature: degrees Fahrenheit (°F)

The SI system

The strength of the SI system is its coherence. There are four mechanical and two electrical base units from which all other quantities are derived. The mechanical ones are:

Length: metre (m)
Mass: kilogram (kg)
Time: second (s)
Temperature: Kelvin (K) or, more commonly, degrees Celsius or Centigrade (°C)

Other units are derived from these: for example the Newton (N) is defined as N = kg m/s^2.

SI prefixes

As a rule, prefixes are generally applied to the basic SI unit. The exception is weight, where the prefix is used with the unit gram (g), rather than the basic SI unit kilogram (kg). Prefixes are not used for units of angular measurement (degrees, radians), time (seconds) or temperature (°C or K).

Prefixes are generally chosen in such a way that the numerical value of a unit lies between 0.1 and 1000 (see Table 1.2). For example

28 kN	rather than	2.8×10^4 N
1.25 mm	rather than	0.00125 m
9.3 kPa	rather than	9300 Pa

Table 1.2 SI unit prefixes

Multiplication factor			*Prefix*	*Symbol*
1 000 000 000 000 000 000 000 000	=	10^{24}	yotta	Y
1 000 000 000 000 000 000 000	=	10^{21}	zetta	Z
1 000 000 000 000 000 000	=	10^{18}	exa	E
1 000 000 000 000 000	=	10^{15}	peta	P
1 000 000 000 000	=	10^{12}	tera	T
1 000 000 000	=	10^{9}	giga	G
1 000 000	=	10^{6}	mega	M
1 000	=	10^{3}	kilo	k
100	=	10^{2}	hicto	h
10	=	10^{1}	deka	da
0.1	=	10^{-1}	deci	d
0.01	=	10^{-2}	centi	c
0.001	=	10^{-3}	milli	m
0.000 001	=	10^{-6}	micro	μ
0.000 000 001	=	10^{-9}	nano	n
0.000 000 000 001	=	10^{-12}	pico	p
0.000 000 000 000 001	=	10^{-15}	femto	f
0.000 000 000 000 000 001	=	10^{-18}	atto	a
0.000 000 000 000 000 000 001	=	10^{-21}	zepto	z
0.000 000 000 000 000 000 000 001	=	10^{-24}	yocto	y

1.3 Conversions

Units often need to be converted. The least confusing way to do this is by expressing equality. For example, to convert 600 lb to kilograms (kg), using 1 kg = 2.205 lb

Add denominators as

$$\frac{1\text{ kg}}{x} = \frac{2.205\text{ lb}}{600\text{ lb}}$$

Solve for x

$$x = \frac{600 \times 1}{2.205} = 272.1\text{ kg}$$

Hence 600 lb = 272.1 kg

Setting out calculations in this way can help avoid confusion, particularly when they involve large numbers and/or several sequential stages of conversion.

Force or thrust

Force and thrust (see Table 1.3) are important quantities in determining the stresses in a mechanical body. Both SI and imperial units are in common use.

Table 1.3 Force (*F*) or thrust

Unit	*lbf*	*gf*	*kgf*	*N*
1 pound-thrust (lbf)	1	453.6	0.4536	4.448
1 gram-force (gf)	2.205×10^{-3}	1	0.001	9.807×10^{-3}
1 kilogram-force (kgf)	2.205	1000	1	9.807
1 Newton (N)	0.2248	102.0	0.1020	1

Note: Strictly, all the units in the table, except the Newton (N), represent weight equivalents of mass and so depend on the 'standard' acceleration due to gravity (*g*). The true SI unit of force is the Newton (N), which is equivalent to 1 kgm/s^2.

Weight

The true weight of a body is a measure of the gravitational attraction of the earth on it. Since this attraction is a force, the weight of a body is correctly expressed in Newtons.

Force (N) = mass (kg) × g(m/s²)
1 ton (US) = 2000 lb = 907.2 kg
1 tonne (metric) = 1000 kg = 2205 lb

Density

Density is defined as mass per unit volume. Table 1.4 shows the conversions between units.

Table 1.4 Density (ρ)

Unit	*lb/in³*	*lb/ft³*	*kg/m³*	*g/cm³*
1 lb per in³	1	1728	2.768×10^{4}	27.68
1 lb per ft³	5.787×10^{-4}	1	16.02	1.602×10^{-2}
1 kg per m³	3.613×10^{-5}	6.243×10^{-2}	1	0.001
1 g per cm³	3613×10^{-2}	62.43	1000	1

Pressure

1 Pascal (Pa) = 1 N/m²
1 Pa = $1.450\ 38 \times 10^{-4}$ lbf/in²

In practice, pressures in SI units are measured in MPa, bar, atmospheres, Torr, or the height of a liquid column, depending on the application. See Figs 1.1 and 1.2 and Table 1.5.

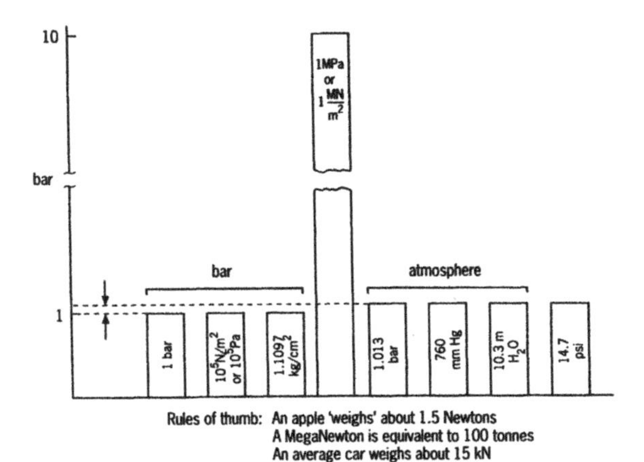

Fig. 1.1 Pressure equivalents

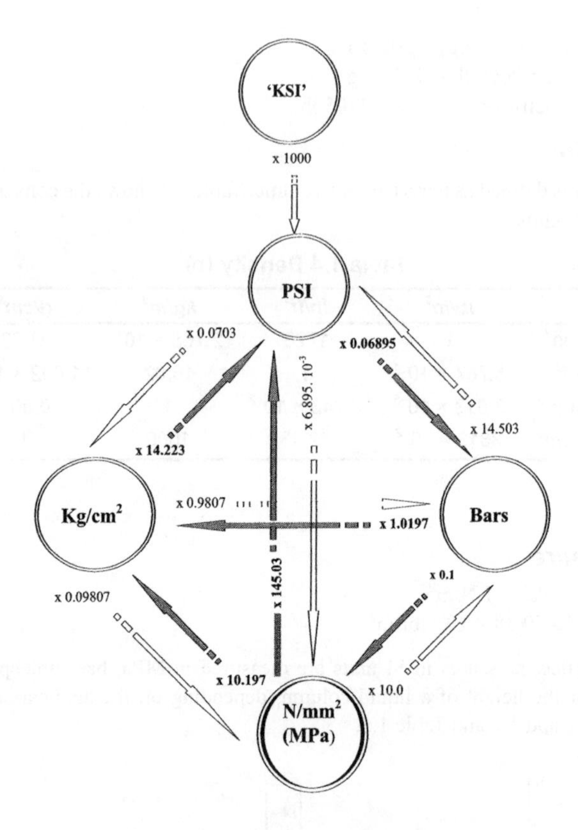

Fig. 1.2 Pressure conversions

Table 1.5 Pressure (p)

Unit	*lb/in^2 (psi)*	*lb/ft^2*	*atm*	*in H_2O*	*cm Hg*	*N/m^2 (Pa)*
1 lb per in^2 (psi)	1	144	6.805×10^{-2}	27.68	5.171	6.895×10^{3}
1 lb per ft^2	6.944×10^{-3}	1	4.725×10^{-4}	0.1922	3.591×10^{-2}	47.88
1 atmosphere (atm)	14.70	2116	1	406.8	76	1.013×10^{5}
1 in of water at 39.2 °F (4 °C)	3.613×10^{-2}	5.02	2.458×10^{-3}	1	0.1868	249.1
1 cm of mercury at 32 °F (0 °C)	0.1934	27.85	1.316×10^{-2}	5.353	1	1333
1 N per m^2 (Pa)	1.450×10^{-4}	2.089×10^{-2}	9.869×10^{-6}	4.015×10^{-3}	7.501×10^{-4}	1

So, for liquid columns

1 in H_2O = 25.4 mm H_2O = 249.089 Pa
1 in Hg = 13.59 in H_2O = 3385.12 Pa = 33.85 mbar
1 mm Hg = 13.59 mm H_2O = 133.3224 Pa = 1.333 224 mbar
1 mm H_2O = 9.806 65 Pa
1 Torr = 133.3224 Pa

For conversion of liquid column pressures: 1 in = 25.4 mm

Temperature

The basic USCS unit of temperature is degrees Fahrenheit (°F). The SI unit is Kelvin (K). The most commonly used unit is degrees Celsius (°C).

Absolute zero is defined as 0 K or –273.15 °C, the point at which a perfect gas has zero volume. See Figs 1.3 and 1.4.

$$°C = \frac{5}{9}(°F - 32)$$
$$°F = \frac{9}{5}(°C) + 32$$

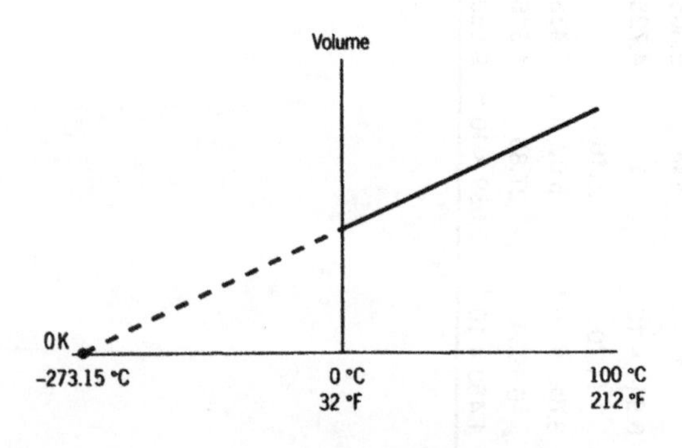

Fig. 1.3 Temperature

Heat and work

The basic unit for heat 'energy' is the Joule.

Specific heat 'energy' is measured in Joules per kilogram (J/kg) in SI units and BTU/lb in USCS units.

1 J/kg = 0.429 923 × 10^{-3} BTU/lb

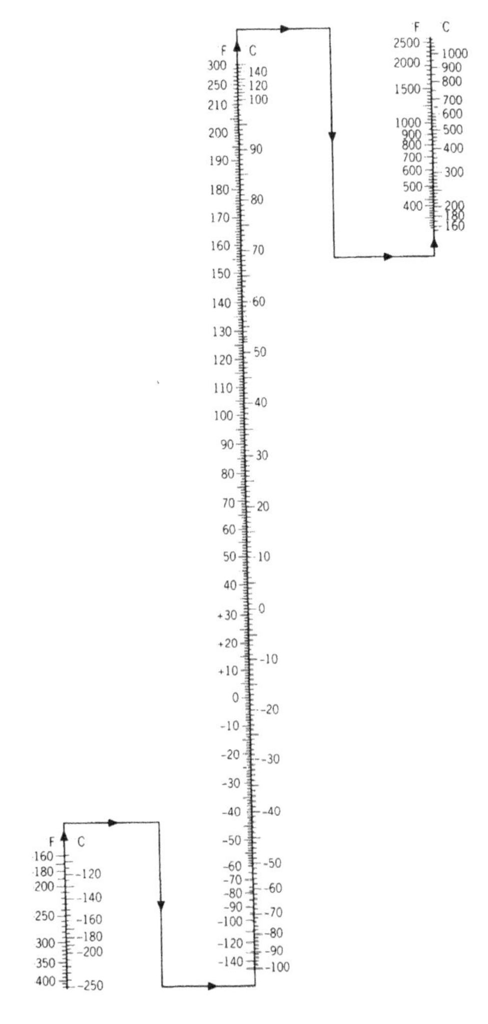

Fig. 1.4 Temperature conversion

Table 1.6 shows common conversions.

Specific heat is measured in BTU/lb °F in USCS units [or in SI; Joules per kilogram Kelvin (J/kg K)].

1 BTU/lb °F = 4186.798 J/kg K

1 J/kg K = 0.238 846 × 10^{-3} BTU/lb °F

1 kcal/kg K = 4186.8 J/kg K

Table 1.6 Heat

	BTU	*ft-lb*	*hp-h*	*cal*	*J*	*kW-h*
1 British thermal unit (BTU)	1	777.9	3.929×10^{-4}	252	1055	2.93×10^{-4}
1 foot-pound (ft-lb)	1.285×10^{-3}	1	5.051×10^{-7}	0.3239	1.356	3.766×10^{-7}
1 horsepower-hour (hp-h)	2545	1.98×10^{6}	1	6.414×10^{5}	2.685×10^{6}	0.7457
1 calorie (cal)	3.968×10^{-3}	3.087	1.559×10^{-6}	1	4.187	1.163×10^{-6}
1 Joule (J)	9.481×10^{-4}	0.7376	3.725×10^{-7}	0.2389	1	2.778×10^{-7}
1 kilowatt hour (kW-h)	3413	2.655×10^{6}	1.341	8.601×10^{5}	3.6×10^{6}	1

Heat flowrate is also defined as power, with the USCS unit of BTU/h [or in SI, in Watts (W)].

1 BTU/h = 0.07 cal/s = 0.293 W
1 W = 3.412 14 BTU/h = 0.238 846 cal/s

Power

BTU/h or horsepower (hp) are normally used in USCS or, in SI, kilowatts (kW). See Table 1.7.

Flow

The basic unit of volume flowrate in SI is litre/s. In the USA it is US gal/min.

1 US gallon = 4 quarts = 128 US fluid ounces = 231 in^3
1 US gallon = 0.8 British imperial gallons = 3.788 33 litres
1 US gallon/minute = 6.314 01 × 10^{-5} m^3/s = 0.2273 m^3/h
1 m^3/s = 1000 litre/s
1 litre/s = 2.12 ft^3/min

Torque

The basic USCS unit of torque is the foot pound (ft.lbf) – in SI it is the Newton metre (Nm). You may also see this referred to as 'moment of force' (see Fig. 1.5).

1 ft.lbf = 1.357 Nm
1 kgf.m = 9.81 Nm

Stress

In SI the basic unit of stress is the Pascal (Pa). One Pascal is an impractical small unit so MPa is normally used (see Fig. 1.6). In the USCS system, stress is measured in lb/in^2 – the same unit used for pressure, although it is a different physical quantity.

1 lb/in^2 = 6895 Pa
1 MPa = 1 MN/m^2 = 1 N/mm^2
1 kgf/mm^2 = 9.806 65 MPa

Table 1.7 Power (*P*)

	BTU/h	*BTU/s*	*ft-lb/s*	*hp*	*cal/s*	*kW*	*W*
1 BTU/h	1	2.778×10^{-4}	0.2161	3.929×10^{-4}	7.000×10^{-2}	2.930×10^{-4}	0.2930
1 BTU/s	3600	1	777.9	1.414	252.0	1.055	1.055×10^{-3}
1 ft–lb/s	4.628	1.286×10^{-3}	1	1.818×10^{-3}	0.3239	1.356×10^{-3}	1.356
1 hp	2545	0.7069	550	1	178.2	0.7457	745.7
1 cal/s	14.29	0.3950	3.087	5.613×10^{-3}	1	4.186×10^{-3}	4.186
1 kW	3413	0.9481	737.6	1.341	238.9	1	1000
1 W	3.413	9.481×10^{-4}	0.7376	1.341×10^{-3}	0.2389	0.001	1

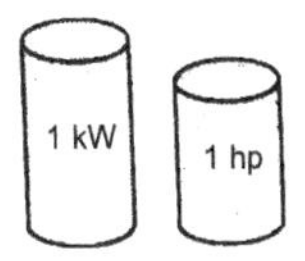

Fig. 1.5 Torque

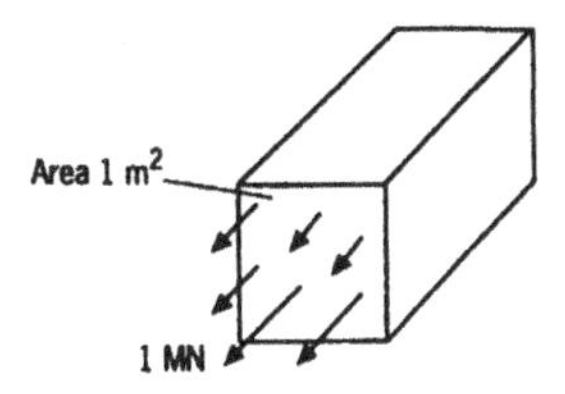

Fig. 1.6 Stress

Linear velocity (speed)

Linear velocity (see Table 1.8) is an important quantity in determining kinetic forces in a component. The basic USCS unit for linear velocity is feet per second (in SI it is m/s).

Table 1.8 Velocity (v)

Item	*ft/s*	*km/h*	*m/s*	*mile/h*	*cm/s*
1 ft per s	1	1.097	0.3048	0.6818	30.48
1 km per h	0.9113	1	0.2778	0.6214	27.78
1 m per s	3.281	3.600	1	2.237	100
1 mile per h	1.467	1.609	0.4470	1	44.70
1 cm per s	3.281×10^{-2}	3.600×10^{-2}	0.0100	2.237×10^{-2}	1

Acceleration

The basic unit of acceleration in SI is m/s^2. The USCS unit is feet per second squared (ft/s^2).

$1\ \text{ft/s}^2 = 0.3048\ \text{m/s}^2$
$1\ \text{m/s}^2 = 3.280\ 84\ \text{ft/s}^2$

Standard gravity (g) is normally taken as 9.806 65 m/s^2 (32.1740 ft/s^2).

Angular velocity

The basic unit of angular velocity is radians per second (rad/s).

1 rad/s = 0.159 155 rev/s = 57.2958 degree/s

The radian is also the SI unit used for plane angles.

- A complete circle is 2π radians (see Fig. 1.7).
- A quarter-circle (90°) is $\pi/2$ or 1.57 radians.
- One degree = $\pi/180$ radians.

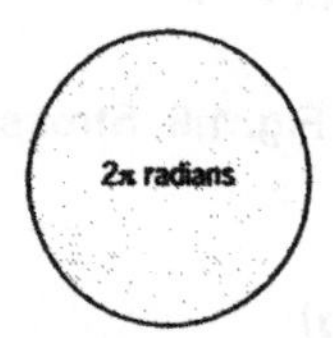

Fig. 1.7 Angular measure

Length and area

Both SI and imperial units are in common use. Table 1.9 shows the conversion.

Comparative lengths in USCS and SI units are:

1 ft = 0.3048 m
1 in = 25.4 mm

Small dimensions are measured in 'micromeasurements' (see Fig. 1.8).

Table 1.9 Area (*A*)

Unit	*sq in*	*sq ft*	*sq yd*	*sq mile*	cm^2	dm^2	m^2	*a*	*ha*	km^2
1 square inch	1	–	–	–	6.452	0.064 52	–	–	–	–
1 square foot	144	1	0.1111	–	929	9.29	0.0929	–	–	–
1 square yard	1296	9	1	–	8361	83.61	0.8361	–	–	–
1 square mile	–	–	–	1	–	–	–	–	259	2.59
1 cm^2	0.155	–	–	–	1	0.01	–	–	–	–
1 dm^2	15.5	0.1076	0.011 96	–	100	1	0.01	–	–	–
1 m^2	1550	10.76	1.196	–	10 000	100	1	0.01	–	–
1 are (a)	–	1076	119.6	–	–	10 000	100	1	0.01	–
1 hectare (ha)	–	–	–	–	–	–	10 000	100	1	0.01
1 km^2	–	–	–	0.386 1	–	–	–	10 000	100	1

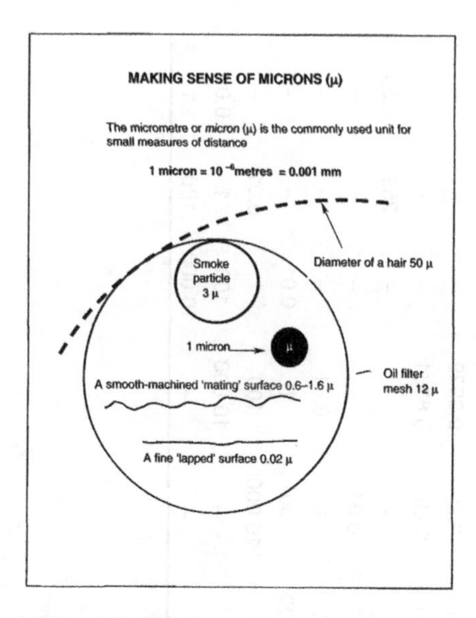

Fig. 1.8 Making sense of microns

Viscosity

Dynamic viscosity (μ) is measured in the SI system in Ns/m^2 or Pascal seconds (Pa s). In the USCS system it is lbf.s/ft^2.

1 lbf.s/ft^2 = 4.882 kgf.s/m^2 = 4.882 Pa s
1 Pa s = 1 N s/m^2 = 1 kg/m s

A common unit of dynamic viscosity is the centipoise (cP). See Table 1.10.

Table 1.10 Dynamic viscosity (μ)

Unit	*lbf-s/ft^2*	*Centipoise*	*Poise*	*kgf/ms*
1 lb (force)–s per ft^2	1	4.788×10^4	4.788×10^2	4.882
1 Centipoise	2.089×10^{-5}	1	10^{-2}	1.020×10^{-4}
1 Poise	2.089×10^{-3}	100	1	1.020×10^{-2}
1 N–s per m^2	0.2048	9.807×10^3	98.07	1

- Kinematic viscosity (ν) is a function of dynamic viscosity.
- Kinematic viscosity = dynamic viscosity/density, i.e. $\nu = \mu/\rho$.

Units such as Saybolt Seconds Universal (SSU) and Stokes (St) are used. The USCS unit is ft^2/s.

$1\ m^2/s = 10.7639\ ft^2/s = 5.580\ 01 \times 10^6\ in^2/h$
1 Stoke (St) $= 100$ centistokes (cSt) $= 10^{-4}\ m^2/s$
$1\ St \cong 0.002\ 26\ (SSU) - 1.95/(SSU)$ for $32 < SSU < 100$ s
$1\ St \cong 0.002\ 20\ (SSU) - 1.35/(SSU)$ for $SSU > 100$ s

1.4 Consistency of units

Within any system of units, the consistency of units forms a 'quick check' of the validity of equations. The units must match on both sides.

Example: (in USCS units)

To check kinematic viscosity (ν) = $\dfrac{\text{dynamic viscosity } (\mu)}{\text{density } (\rho)} = \mu \times 1/\rho$

$$\frac{ft^2}{s} = \frac{lbf.s}{ft^2} \times \frac{ft^4}{lbf.s^2}$$

Cancelling gives $$\frac{ft^2}{s} = \frac{s.ft^4}{s^2.ft^2} = \frac{ft^2}{s}$$

OK, units match.

1.5 Foolproof conversions: using unity brackets

When converting between units it is easy to make mistakes by dividing by a conversion factor instead of multiplying, or vice versa. The best way to avoid this is by using the technique of unity brackets.

A unity bracket is a term, consisting of a numerator and denominator in different units, which has a value of unity.

For example

$$\left[\frac{2.205\ lb}{kg}\right]$$

or

$$\left[\frac{kg}{2.205\ lb}\right]$$

are unity brackets, as are

$$\left[\frac{25.4\ \text{mm}}{\text{in}}\right]$$

or

$$\left[\frac{\text{in}}{25.4\ \text{mm}}\right]$$

or

$$\left[\frac{\text{Atmosphere}}{101\ 325\ \text{Pa}}\right]$$

As the value of the term inside the bracket is unity, it has no effect on any term that it multiplies.

Example: Convert the density of titanium 6 Al 4 V; $\rho = 0.16\ \text{lb/in}^3$ to kg/m^3

Step 1

State the initial value

$$\rho = \frac{0.16\ \text{lb}}{\text{in}^3}$$

Step 2

Apply the 'weight' unity bracket

$$\rho = \frac{0.16\ \text{lb}}{\text{in}^3}\left[\frac{\text{kg}}{2.205\ \text{lb}}\right]$$

Step 3

Apply the 'dimension' unity brackets (cubed)

$$\rho = \frac{0.16\ \text{lb}}{\text{in}^3}\left[\frac{\text{kg}}{2.205\ \text{lb}}\right]\left[\frac{\text{in}}{25.4\ \text{mm}}\right]^3\left[\frac{1000\ \text{mm}}{\text{m}}\right]^3$$

Step 4

Expand and cancel*

$$\rho = \frac{0.16\ \cancel{\text{lb}}}{\cancel{\text{in}^3}}\left[\frac{\text{kg}}{2.205\ \cancel{\text{lb}}}\right]\left[\frac{\cancel{\text{in}^3}}{(25.4)^3\ \cancel{\text{mm}^3}}\right]\left[\frac{(1000)^3\ \cancel{\text{mm}^3}}{\text{m}^3}\right]$$

$$\rho = \frac{0.16\ \text{kg}\ (1000)^3}{2.205\ (25.4)^3\ \text{m}^3}$$

$\rho = 4428.02\ \text{kg/m}^3$: Answer

* Take care to use the correct algebraic rules for the expansion, for example

$$(a.b)^N = a^N.b^N$$

not

$$a.b^N$$

And

$$\left[\frac{1000\ \text{mm}}{\text{m}}\right]^3$$

expands to

$$\frac{(1000)^3.(\text{mm})^3}{(\text{m})^3}$$

Unity brackets can be used for all units conversions provided you follow the rules for algebra correctly.

1.6 Imperial–metric conversions

Conversions from metric to imperial units, and vice versa, often use rounding to a prescribed number of significant figures. Table 1.11 shows a conversion in common use.

Table 1.11 Imperial–metric conversions

Fraction (in)	*Decimal (in)*	*Millimetre (mm)*
1/64	0.01562	0.39687
1/32	0.03125	0.79375
3/64	0.04687	1.19062
1/16	0.06250	1.58750
5/64	0.07812	1.98437
3/32	0.09375	2.38125
7/64	0.10937	2.77812
1/8	0.12500	3.17500
9/64	0.14062	3.57187
5/32	0.15625	3.96875
11/64	0.17187	4.36562
3/16	0.18750	4.76250
13/64	0.20312	5.15937
7/32	0.21875	5.55625
15/64	0.23437	5.95312

Table 1.11 Cont.

1/4	0.25000	6.35000
17/64	0.26562	6.74687
9/32	0.28125	7.14375
19/64	0.29687	5.54062
15/16	0.31250	7.93750
21/64	0.32812	8.33437
11/32	0.34375	8.73125
23/64	0.35937	9.12812
3/8	0.37500	9.52500
25/64	0.39062	9.92187
13/32	0.40625	10.31875
27/64	0.42187	10.71562
7/16	0.43750	11.11250
29/64	0.45312	11.50937
15/32	0.46875	11.90625
31/64	0.48437	12.30312
1/2	0.50000	12.70000
33/64	0.51562	13.09687
17/32	0.53125	13.49375
35/64	0.54687	13.89062
9/16	0.56250	14.28750
37/64	0.57812	14.68437
19/32	0.59375	15.08125
39/64	0.60937	15.47812
5/8	0.62500	15.87500
41/64	0.64062	16.27187
21/32	0.65625	16.66875
43/64	0.67187	17.06562
11/16	0.68750	17.46250
45/64	0.70312	17.85937
23/32	0.71875	18.25625
47/64	0.73437	18.65312
3/4	0.75000	19.05000
49/64	0.76562	19.44687
25/32	0.78125	19.84375
51/64	0.79687	20.24062

Table 1.11 Cont.

13/16	0.81250	20.63750
53/64	0.82812	21.03437
27/32	0.84375	21.43125
55/64	0.85937	21.82812
7/8	0.87500	22.22500
57/64	0.89062	22.62187
29/32	0.90625	23.01875
59/64	0.92187	23.41562
15/16	0.93750	23.81250
61/64	0.95312	24.20937
31/32	0.96875	24.60625
63/64	0.98437	25.00312
1	1.00000	25.40000

1.7 Dimensional analysis

Dimensional analysis (DA) – what is it?

Dimensional analysis is a technique based on the idea that one physical quantity is related to others in a precise mathematical way. It is used in rotating and hydraulic equipment design for:

- checking the validity of equations;
- finding the arrangement of variables in a formula;
- helping to tackle problems that do not possess a complete theoretical solution, particularly those involving fluid mechanics.

Primary and secondary quantities

Primary quantities are quantities that are absolutely independent of each other. They are

M Mass
L Length
T Time

For example: Velocity (v) is represented by length divided by time, and this is shown by

$[v] = \frac{\mathrm{L}}{\mathrm{T}}$: note the square brackets denoting 'the dimension of'.

Table 1.12 shows the most commonly used quantities.

Table 1.12 Dimensional analysis quantities

Quantity	*Dimensions*
Mass, *m*	M
Length, *l*	L
Time, *t*	T
Area, *a*	L^2
Volume, *V*	L^3
First moment of area	L^3
Second moment of area	L^4
Velocity, *v*	LT^{-1}
Acceleration, *a*	LT^{-2}
Angular velocity, ω	T^{-1}
Angular acceleration, α	T^{-2}
Frequency, *f*	T^{-1}
Force, *F*	MLT^{-2}
Stress {Pressure}, *S* {*P*}	$ML^{-1}T^{-2}$
Torque, *T*	ML^2T^{-2}
Modulus of elasticity, *E*	$ML^{-1}T^{-2}$
Work, *W*	ML^2T^{-2}
Power, *P*	ML^2T^{-3}
Density, ρ	ML^{-3}
Dynamic viscosity, μ	$ML^{-1}T^{-1}$
Kinematic viscosity, ν	L^2T^{-1}

Hence velocity is termed a *secondary quantity* because it can be expressed in terms of primary quantities.

An example of deriving formulae using DA

To find the frequencies n of eddies behind a cylinder situated in a free stream of pumped fluid, we can assume that n is related in some way to the diameter d of the cylinder, the speed V of the fluid stream, the fluid density ρ, and the kinematic viscosity ν of the fluid.
i.e.

$$n = \phi\{d, V, \rho, \nu\}$$

Introducing a numerical constant Y and some possible exponentials gives

$$n = \text{Y}\,\{d^a, V^b, \rho^c, v^d\}$$

Y is a dimensionless constant so, in dimensional analysis terms, this equation becomes, after substituting primary dimensions

$$\begin{aligned} \text{T}^{-1} &= \text{L}^a(\text{LT}^{-1})^b\,(\text{ML}^{-3})^c\,(\text{L}^2\text{T}^{-1})^d \\ &= \text{L}^a\,\text{L}^b\,\text{T}^{-b}\,\text{M}^c\,\text{L}^{-3c}\,\text{L}^{2d}\,\text{T}^{-d} \end{aligned}$$

In order for the equation to balance

For M
c must $= 0$

For L
$a + b - 3c + 2d = 0$

For T
$-b - d = -1$

Solving for a, b, c in terms of d gives

$a = -1 - d$
$b = 1 - d$

Giving

$$n = d^{(-1-d)}\,V^{(1-d)}\,\rho^0\,v^d$$

Rearranging gives

$$nd/V = (Vd/v)\text{X}$$

Note how dimensional analysis can give the 'form' of the formula but not the numerical value of the undetermined constant X which, in this case, is a compound constant containing the original constant Y and the unknown index d.

1.8 Essential mathematics

Basic algebra

$a^m \times a^n = a^{m+n}$
$a^m \div a^n = a^{m-n}$
$(a^m)^n = a^{mn}$
$\sqrt[n]{a^m} = a^{m/n}$

$$\frac{1}{a^n} = a^{-n}$$

$a^0 = 1$
$(a^n b^m)p = a^{np} b^{mp}$

$$\left(\frac{a}{b}\right)^n = \frac{a^n}{b^n}$$

$^n\sqrt{(ab)} = {}^n\sqrt{a} \times {}^n\sqrt{b}$

$$^n\sqrt{\frac{a}{b}} = \frac{^n\sqrt{a}}{^n\sqrt{b}}$$

Logarithms

If $N = a^x$ then $\log_a N = x$ and $N = a^{\log_a N}$

$$\log_a N = \frac{\log_b N}{\log_b a}$$

$\log (ab) = \log a + \log b$

$$\log\left(\frac{a}{b}\right) = \log a - \log b$$

$\log a^n = n \log a$

$$\log {}^n\sqrt{a} = \frac{1}{n} \log a$$

$\log_a 1 = 0$
$\log_e N = 2.3026 \log_{10} N$

Quadratic equations

If $ax^2 + bx + c = 0$

$$x = \frac{-b \pm \sqrt{(b^2 - 4ac)}}{2a}$$

If $b^2 - 4ac > 0$ the equation $ax^2 + bx + c = 0$ yields two real and different roots.
If $b^2 - 4ac = 0$ the equation $ax^2 + bx + c = 0$ yields coincident roots.
If $b^2 - 4ac < 0$ the equation $ax^2 + bx + c = 0$ has complex roots.
If α and β are the roots of the equation $ax^2 + bx + c = 0$ then

sum of the roots $= \alpha + \beta = -\frac{b}{a}$ product of the roots $= \alpha\beta = \frac{c}{a}$

The equation whose roots are α and β is $x^2 - (\alpha + \beta)x + \alpha\beta = 0$.
Any quadratic function $ax^2 + bx + c$ can be expressed in the form $p(x + q)^2 + r$ or $r - p(x + q)^2$, where r, p, and q are all constants.

The function $ax^2 + bx + c$ will have a maximum value if a is negative and a minimum value if a is positive.
If $ax^2 + bx + c = p(x + q)^2 + r = 0$ the minimum value of the function occurs when $(x + q) = 0$ and its value is r.
If $ax^2 + bx + c = r - p(x + q)^2$ the maximum value of the function occurs when $(x + q) = 0$ and its value is r.

Cubic equations

$x^3 + px^2 + qx + r = 0$

$x = y - \frac{1}{3}p$ gives $y^3 + 3ay + 2b = 0$

where

$$3a = -q - \frac{1}{3}p^2, \quad 2b = \frac{2}{27}p^3 - \frac{1}{3}pq + r$$

On setting

$$S = [-b + (b^2 + a^3)^{1/2}]^{1/3} \text{ and } T = [-b - (b^2 + a^3)^{1/2}]^{1/3}$$

the three roots are

$$x_1 = S + T - \frac{1}{3}p$$

$$x_2 = -\frac{1}{2}(S + T) + \frac{\sqrt{3}}{2}i(S - T) - \frac{1}{3}p$$

$$x_3 = -\frac{1}{2}(S + T) - \frac{\sqrt{3}}{2}i(S - T) - \frac{1}{3}p.$$

For real coefficients
all roots are real if $b^2 + a^3 \le 0$,
one root is real if $b^2 + a^3 > 0$.
At least two roots are equal if $b^2 + a^3 = 0$
Three roots are equal if $a = 0$ and $b = 0$.
For $b^2 + a^3 < 0$ there are alternative expressions

$$x_1 = 2c\cos\frac{1}{3}\theta - \frac{1}{3}p$$

$$x_2 = 2c\cos\frac{1}{3}(\theta + 2\pi) - \frac{1}{3}p$$

$$x_3 = 2c\cos\frac{1}{3}(\theta + 4\pi) - \frac{1}{3}p$$

where $c^2 = -a$ and $\cos\theta = -\frac{b}{c^3}$

Complex numbers

If x and y are real numbers and $i = \sqrt{-1}$ then the complex number $z = x + iy$ consists of real part x and the imaginary part iy.
$\bar{z} = x - iy$ is the conjugate of the complex number $z = x + iy$.
If $x + iy = a + ib$ then $x = a$ and $y = b$

$$(a + ib) + (c + id) = (a + c) = i(b + d)$$

$$(a + ib) - (c + id) = (a - c) + i(b - d)$$

$$(a + ib)(c + id) = (ac - bd) + i(ad + bc)$$

$$\frac{a+ib}{c+id} = \frac{ac+bd}{c^2+d^2} + i\frac{bc-ad}{c^2+d^2}$$

Every complex number may be written in polar form. Thus

$$x + iy = r(\cos\theta + i\sin\theta) = r\angle\theta$$

r is called the *modulus* of z and this may be written $r = |z|$

$$r = \sqrt{(x^2 + y^2)}$$

θ is called the *argument* and this may be written $\theta = \arg z$

$$\tan\theta = \frac{y}{x}$$

If $z_1 = r(\cos\theta_1 + i\sin\theta_1)$ and $z_2 = r_2(\cos\theta_2 + i\sin\theta_2)$

$$z_1 z_2 = r_1 r_2[\cos(\theta_1 + \theta_2) + i\sin(\theta_1 + \theta_2)] = r_1 r_2\angle(\theta_1 + \theta_2)$$

$$\frac{z_1}{z_2} = \frac{r_1[\cos(\theta_1 - \theta_2) + i\sin(\theta_1 + \theta_2)]}{r_2} = \frac{r_1}{r_2}\angle(\theta_1 - \theta_2)$$

Standard series

Binomial series

$$(a + x)^n = a^n + na^{n-1}x + \frac{n(n-1)}{2!}a^{n-2}x^2 + \frac{n(n-1)(n-2)}{3!}a^{n-3}x^3 + \ldots \; (x^2 < a^2)$$

The number of terms becomes infinite when n is negative or fractional.

$$(a - bx)^{-1} = \frac{1}{a}\left(1 + \frac{bx}{a} + \frac{b^2x^2}{a^2} + \frac{b^3x^3}{a^3} + \ldots\right) \quad (b^2x^2 < a^2)$$

Exponential series

$$a^x = 1 + x \ln a + \frac{(x \ln a)^2}{2!} + \frac{(x \ln a)^3}{3!} + \ \ldots$$

$$\mathrm{e}^x = 1 + x + \frac{x^2}{2!} + \frac{x^3}{3!} + \ \ldots$$

Logarithmic series

$$\ln x = (x-1) - \frac{1}{2}(x-1)^2 + \frac{1}{3}(x-1)^3 - \ \ldots \quad (0 < x < 2)$$

$$\ln x = \frac{x-1}{x} + \frac{1}{2}\left(\frac{x-1}{x}\right)^2 + \frac{1}{3}\left(\frac{x-1}{x}\right)^3 + \ \ldots \quad \left(x > \frac{1}{2}\right)$$

$$\ln x = 2\left[\frac{x-1}{x+1} . \frac{1}{3}\left(\frac{x-1}{x+1}\right)^3 + \frac{1}{5}\left(\frac{x-1}{x+1}\right)^5 + \ \ldots\right] \quad (x \text{ positive})$$

$$\ln\ (1+x) = x - \frac{x^2}{2} + \frac{x^3}{3} - \frac{x^4}{4} + \ \ldots$$

Trigonometric series

$$\sin x = x - \frac{x^3}{3!} + \frac{x^5}{5!} - \frac{x^7}{7!} + \ \ldots$$

$$\cos x = 1 - \frac{x^2}{2!} + \frac{x^4}{4!} - \frac{x^6}{6!} + \ \ldots$$

$$\tan x = x + \frac{x^3}{3} + \frac{2x^5}{15} + \frac{17x^7}{315} + \frac{62x^9}{2835} + \ \ldots \quad \left(x^2 < \frac{\pi^2}{4}\right)$$

$$\sin^{-1} x = x + \frac{1}{2}\frac{x^3}{3} + \frac{1 \cdot 3}{2 \cdot 4}\frac{x^5}{5} + \frac{1 \cdot 3 \cdot 5}{2 \cdot 4 \cdot 6}\frac{x^7}{7} + \ \ldots \quad (x^2 < 1)$$

$$\tan^{-1} x = x - \frac{1}{3}x^3 + \frac{1}{5}x^5 - \frac{1}{7}x^7 + \ \ldots \quad (x^2 \leqq 1)$$

Vector algebra

Vectors have direction and magnitude and satisfy the triangle rule for addition. Quantities such as velocity, force, and straight-line displacements may be represented by vectors. Three-dimensional vectors are used to represent physical quantities in space, for example A_x, A_y, A_z or $A_x\boldsymbol{i} + A_y\boldsymbol{j} + A_z\boldsymbol{k}$.

Vector addition

The vector sum $\boldsymbol{V}$ of any number of vectors $\boldsymbol{V}_1$, $\boldsymbol{V}_2$, $\boldsymbol{V}_3$ where $\boldsymbol{V}_1 = a_1\boldsymbol{i} + b_1\boldsymbol{j} + c_1\boldsymbol{k}$, etc., is given by

$$\boldsymbol{V} = \boldsymbol{V}_1 + \boldsymbol{V}_2 + \boldsymbol{V}_3 + \ldots = (a_1 + a_2 + a_3 + \ldots)\boldsymbol{i}$$

$$+(b_1 + b_2 + b_3 + \ldots)\boldsymbol{j} + (c_1 + c_2 + c_3 + \ldots)\boldsymbol{k}$$

Product of a vector V by a scalar quantity s

$$s\boldsymbol{V} = (sa)\boldsymbol{i} + (sb)\boldsymbol{j} + (sc)\boldsymbol{k}$$

$$(s_1 + s_2)\boldsymbol{V} = s_1\boldsymbol{V} + s_2\boldsymbol{V} \qquad (\boldsymbol{V}_1 + \boldsymbol{V}_2)s = \boldsymbol{V}_1 s + \boldsymbol{V}_2 s$$

where $s\boldsymbol{V}$ has the same direction as $\boldsymbol{V}$, and its magnitude is s times the magnitude of $\boldsymbol{V}$.

Scalar product of two vectors, $V_1 \cdot V_2$

$$\boldsymbol{V}_1 \cdot \boldsymbol{V}_2 = |\boldsymbol{V}_1||\boldsymbol{V}_2|\cos\phi$$

where ϕ is the angle between $\boldsymbol{V}_1$ and $\boldsymbol{V}_2$.

Vector product of two vectors, $V_1 \times V_2$

$$|\boldsymbol{V}_1 \times \boldsymbol{V}_2| = |\boldsymbol{V}_1||\boldsymbol{V}_2|\sin\phi$$

where ϕ is the angle between $\boldsymbol{V}_1$ and $\boldsymbol{V}_2$.

Derivatives of vectors

$$\frac{\mathrm{d}}{\mathrm{d}t}(\boldsymbol{A} \cdot \boldsymbol{B}) = \boldsymbol{A} \cdot \frac{\mathrm{d}\boldsymbol{B}}{\mathrm{d}t} + \boldsymbol{B} \cdot \frac{\mathrm{d}\boldsymbol{A}}{\mathrm{d}t}$$

If $e(t)$ is a unit vector $\frac{\mathrm{d}e}{\mathrm{d}t}$ is perpendicular to e: that is $e \cdot \frac{\mathrm{d}e}{\mathrm{d}t} = 0$.

$$\frac{\mathrm{d}}{\mathrm{d}t}(\boldsymbol{A} \times \boldsymbol{B}) = \boldsymbol{A} \times \frac{\mathrm{d}\boldsymbol{B}}{\mathrm{d}t} + \frac{\mathrm{d}\boldsymbol{A}}{\mathrm{d}t} \times \boldsymbol{B}$$

$$= -\frac{\mathrm{d}}{\mathrm{d}t}(\boldsymbol{B} \times \boldsymbol{A})$$

Gradient

The gradient (grad) of a scalar field $\phi\,(x, y, z)$ is

$$\operatorname{grad}\phi = \nabla\phi = \left(\boldsymbol{i}\frac{\partial}{\partial x} + \boldsymbol{j}\frac{\partial}{\partial y} + \boldsymbol{k}\frac{\partial}{\partial z}\right)\phi$$

$$= \frac{\partial\phi}{\partial x}\boldsymbol{i} + \frac{\partial\phi}{\partial y}\boldsymbol{j} + \frac{\partial\phi}{\partial z}\boldsymbol{k}$$

Divergence

The divergence (div) of a vector $\boldsymbol{V} = \boldsymbol{V}(x, y, z) = \boldsymbol{V}_x(x, y, z)\boldsymbol{i} + \boldsymbol{V}_y(x, y, z)\boldsymbol{j} + \boldsymbol{V}_z(x, y, z)\boldsymbol{k}$

$$\text{div } \boldsymbol{V} = \nabla \cdot \boldsymbol{V} = \frac{\partial V_x}{\partial x} + \frac{\partial V_y}{\partial y} + \frac{\partial V_z}{\partial z}$$

Curl

Curl (rotation) is

$$\text{curl } \boldsymbol{V} = \nabla \times \boldsymbol{V} = \begin{vmatrix} \boldsymbol{i} & \boldsymbol{j} & \boldsymbol{k} \\ \frac{\partial}{\partial x} & \frac{\partial}{\partial y} & \frac{\partial}{\partial z} \\ V_x & V_y & V_z \end{vmatrix} = \left(\frac{\partial V_z}{\partial y} - \frac{\partial V_y}{\partial z}\right)\boldsymbol{i} + \left(\frac{\partial V_x}{\partial z} - \frac{\partial V_z}{\partial x}\right)\boldsymbol{j} + \left(\frac{\partial V_y}{\partial x} - \frac{\partial V_x}{\partial y}\right)\boldsymbol{k}$$

Differentiation

Rules for differentiation: y, u, and v are functions of x; a, b, c, and n are constants.

$$\frac{\mathrm{d}}{\mathrm{d}x}(\mathrm{a}u \pm \mathrm{b}v) = \mathrm{a}\frac{\mathrm{d}u}{\mathrm{d}x} \pm \mathrm{b}\frac{\mathrm{d}v}{\mathrm{d}x}$$

$$\frac{\mathrm{d}(uv)}{\mathrm{d}x} = u\frac{\mathrm{d}v}{\mathrm{d}x} + v\frac{\mathrm{d}u}{\mathrm{d}x}$$

$$\frac{\mathrm{d}}{\mathrm{d}x}\left(\frac{u}{v}\right) = \frac{1}{v}\frac{\mathrm{d}u}{\mathrm{d}x} - \frac{u}{v^2}\frac{\mathrm{d}v}{\mathrm{d}x}$$

$$\frac{\mathrm{d}}{\mathrm{d}x}(u^{\mathrm{n}}) = \mathrm{n}u^{\mathrm{n}-1}\frac{\mathrm{d}u}{\mathrm{d}x}, \qquad \frac{\mathrm{d}}{\mathrm{d}x}\left(\frac{1}{u^{\mathrm{n}}}\right) = -\frac{\mathrm{n}}{u^{\mathrm{n}+1}}\frac{\mathrm{d}u}{\mathrm{d}x}$$

$$\frac{\mathrm{d}u}{\mathrm{d}x} = 1\bigg/\frac{\mathrm{d}x}{\mathrm{d}u}, \qquad \text{if } \frac{\mathrm{d}x}{\mathrm{d}u} \neq 0$$

$$\frac{\mathrm{d}}{\mathrm{d}x}f(u) = f'(u)\frac{\mathrm{d}u}{\mathrm{d}x}$$

$$\frac{\mathrm{d}}{\mathrm{d}x}\int_{\mathrm{a}}^{x} f(t)\mathrm{d}t = f(x)$$

$$\frac{\mathrm{d}}{\mathrm{d}x}\int_{x}^{\mathrm{b}} f(t)\mathrm{d}t = -f(x)$$

$$\frac{\mathrm{d}}{\mathrm{d}x}\int_{\mathrm{a}}^{\mathrm{b}} f(x,t)\mathrm{d}t = \int_{\mathrm{a}}^{\mathrm{b}} \frac{\partial f}{\partial x}\mathrm{d}t$$

$$\frac{\mathrm{d}}{\mathrm{d}x}\int_{u}^{v} f(x,t)\mathrm{d}t = \int_{v}^{u} \frac{\partial f}{\partial x}\mathrm{d}t + f(x,v)\frac{\mathrm{d}v}{\mathrm{d}x} - f(x,u)\frac{\mathrm{d}u}{\mathrm{d}x}$$

Higher derivatives

$$\text{Second derivative} = \frac{d}{dx}\left(\frac{dy}{dx}\right) = \frac{d^2 y}{dx^2} = f''(x) = y''$$

$$\frac{d^2}{dx^2} f(u) = f''(u)\left(\frac{du}{dx}\right)^2 + f'(u)\frac{d^2 u}{dx^2}$$

Derivatives of exponentials and logarithms

$$\frac{d}{dx}(ax+b)^n = na(ax+b)^{n-1}$$

$$\frac{d}{dx}e^{ax} = ae^{ax}$$

$$\frac{d}{dx}\ln ax = \frac{1}{x}, \quad ax > 0$$

$$\frac{d}{dx}a^u = a^u \ln a \frac{du}{dx}$$

$$\frac{d}{dx}\log_a u = \log_a e \frac{1}{u}\frac{du}{dx}$$

Derivatives of trigonometric functions in radians

$$\frac{d}{dx}\sin x = \cos x, \qquad \frac{d}{dx}\cos x = -\sin x$$

$$\frac{d}{dx}\tan x = \sec^2 x = 1 + \tan^2 x$$

$$\frac{d}{dx}\cot x = -\operatorname{cosec}^2 x$$

$$\frac{d}{dx}\sec x = \frac{\sin x}{\cos^2 x} = \sec x \tan x$$

$$\frac{d}{dx}\operatorname{cosec} x = -\frac{\cos x}{\sin^2 x} = -\operatorname{cosec} x \cot x$$

$$\frac{d}{dx}\arcsin x = -\frac{d}{dx}\arccos x = \frac{1}{(1-x^2)^{1/2}}$$ for angles in the first quadrant.

Derivatives of hyperbolic functions

$$\frac{\mathrm{d}}{\mathrm{d}x}\sinh x = \cosh x, \qquad \frac{\mathrm{d}}{\mathrm{d}x}\cosh x = \sinh x$$

$$\frac{\mathrm{d}}{\mathrm{d}x}\tanh x = \operatorname{sech}^2 x, \qquad \frac{\mathrm{d}}{\mathrm{d}x}\cosh x = -\operatorname{cosech}^2 x$$

$$\frac{\mathrm{d}}{\mathrm{d}x}(\operatorname{arcsinh} x) = \frac{1}{(x^2+1)^{1/2}}, \qquad \frac{\mathrm{d}}{\mathrm{d}x}(\operatorname{arccosh} x) = \frac{\pm 1}{(x^2-1)^{1/2}}$$

Partial derivatives

Let $f(x, y)$ be a function of the two variables x and y. The partial derivative of f with respect to x, keeping y constant, is

$$\frac{\partial f}{\partial x} = \lim_{h \to 0} \frac{f(x+h, \mathrm{y}) - f(x, \mathrm{y})}{h}$$

Similarly the partial derivative of f with respect to y, keeping x constant, is

$$\frac{\partial f}{\partial y} = \lim_{k \to 0} \frac{f(\mathrm{x}, y+k) - f(\mathrm{x}, y)}{k}$$

Chain rule for partial derivatives

To change variables from (x, y) to (u, v) where $u = u(x, y)$, $v = v(x, y)$, both $x = x(u, v)$ and $y = y(u, v)$ exist and $f(x, y) = f[x(u, v), y(u, v)] = F(u, v)$.

$$\frac{\partial F}{\partial u} = \frac{\partial x}{\partial u}\frac{\partial f}{\partial x} + \frac{\partial y}{\partial u}\frac{\partial f}{\partial y}, \qquad \frac{\partial F}{\partial v} = \frac{\partial x}{\partial v}\frac{\partial f}{\partial v} + \frac{\partial y}{\partial v}\frac{\partial f}{\partial y}$$

$$\frac{\partial f}{\partial x} = \frac{\partial u}{\partial x}\frac{\partial F}{\partial u} + \frac{\partial v}{\partial x}\frac{\partial F}{\partial v}, \qquad \frac{\partial f}{\partial y} = \frac{\partial u}{\partial y}\frac{\partial F}{\partial u} + \frac{\partial v}{\partial y}\frac{\partial F}{\partial v}$$

Integration

f(x)	**F(x) = ∫f(x)dx**
x^{a}	$\frac{x^{\mathrm{a}+1}}{\mathrm{a}+1}$, $\mathrm{a} \neq -1$
x^{-1}	$\ln \lvert x \rvert$
e^{kx}	$\frac{\mathrm{e}^{kx}}{k}$
a^{x}	$\frac{\mathrm{a}^{x}}{\ln \mathrm{a}}$, $\mathrm{a} > 0$, $\mathrm{a} \neq 1$

$\ln x$	$x \ln x - x$
$\sin x$	$-\cos x$
$\cos x$	$\sin x$
$\tan x$	$\ln \lvert\sec x\rvert$
$\cot x$	$\ln \lvert\sin x\rvert$
$\sec x$	$\ln \lvert\sec x + \tan x\rvert = \ln\lvert\tan \tfrac{1}{2}(x + \tfrac{1}{2}\pi)\rvert$
$\operatorname{cosec} x$	$\ln\lvert\tan \tfrac{1}{2} x\rvert$
$\sin^2 x$	$\tfrac{1}{2}(x - \tfrac{1}{2} \sin 2x)$
$\cos^2 x$	$\tfrac{1}{2}(x + \tfrac{1}{2} \sin 2x)$
$\sec^2 x$	$\tan x$
$\sinh x$	$\cosh x$
$\cosh x$	$\sinh x$
$\tanh x$	$\ln \cosh x$
$\operatorname{sech} x$	$2 \arctan e^x$
$\operatorname{cosech} x$	$\ln \lvert\tanh \tfrac{1}{2}x\rvert$
$\operatorname{sech}^2 x$	$\tanh x$
$\dfrac{1}{a^2 + x^2}$	$\dfrac{1}{a}\arctan\dfrac{x}{a}, \quad a \neq 0$
$\dfrac{1}{a^2 - x^2}$	$\begin{cases} -\dfrac{1}{2a}\ln\dfrac{a-x}{a+x}, & a \neq 0 \\ \dfrac{1}{2a}\ln\dfrac{x-a}{x+a}, & a \neq 0 \end{cases}$
$\dfrac{1}{(a^2 - x^2)^{1/2}}$	$\arcsin\dfrac{x}{\lvert a\rvert}, \quad a \neq 0$
$\dfrac{1}{(x^2 - a^2)^{1/2}}$	$\begin{cases} \ln\left[x + (x^2 - a^2)^{1/2}\right] \\ \operatorname{arccosh}\dfrac{x}{a}, & a \neq 0 \end{cases}$

Matrices

A matrix that has an array of $(m \times n)$ numbers arranged in m rows and n columns is called an $(m \times n)$ matrix. It is denoted by

$$\begin{pmatrix} a_{11} & a_{12} & \cdots & a_{1n} \\ a_{21} & a_{22} & \cdots & a_{2n} \\ . & . & \cdots & . \\ . & . & \cdots & . \\ . & . & \cdots & . \\ a_{m1} & a_{m2} & \cdots & a_{mn} \end{pmatrix}$$

Square matrix

This is a matrix having the same number of rows and columns.

$$\begin{pmatrix} a_{11} & a_{12} & a_{13} \\ a_{21} & a_{22} & a_{23} \\ a_{31} & a_{32} & a_{33} \end{pmatrix}$$ is a square matrix of order 3 × 3.

Diagonal matrix

This is a square matrix in which all the elements are zero, except those in the leading diagonal.

$$\begin{pmatrix} a_{11} & 0 & 0 \\ 0 & a_{22} & 0 \\ 0 & 0 & a_{33} \end{pmatrix}$$ is a diagonal matrix of order 3 × 3.

Unit matrix

This is a diagonal matrix with the elements in the leading diagonal all equal to 1. All other elements are 0. The unit matrix is denoted by **I**.

$$\mathbf{I} = \begin{pmatrix} 1 & 0 & 0 \\ 0 & 1 & 0 \\ 0 & 0 & 1 \end{pmatrix}$$

Addition of matrices

Two matrices may be added provided that they are of the same order. This is done by adding the corresponding elements in each matrix.

$$\begin{pmatrix} a_{11} & a_{12} & a_{13} \\ a_{21} & a_{22} & a_{23} \end{pmatrix} + \begin{pmatrix} b_{11} & b_{12} & b_{13} \\ b_{21} & b_{22} & b_{23} \end{pmatrix} = \begin{pmatrix} a_{11}+b_{11} & a_{12}+b_{12} & a_{13}+b_{13} \\ a_{21}+b_{21} & a_{22}+b_{22} & a_{23}+b_{23} \end{pmatrix}$$

Subtraction of matrices

Subtraction is done in a similar way to addition except that the corresponding elements are subtracted.

$$\begin{pmatrix} a_{11} & a_{12} \\ a_{21} & a_{22} \end{pmatrix} - \begin{pmatrix} b_{11} & b_{12} \\ b_{21} & b_{22} \end{pmatrix} = \begin{pmatrix} a_{11}-b_{11} & a_{12}-b_{12} \\ a_{21}-b_{21} & a_{22}-b_{22} \end{pmatrix}$$

Scalar multiplication

A matrix may be multiplied by a number as follows

$$b\begin{pmatrix} a_{11} & a_{12} \\ a_{21} & a_{22} \end{pmatrix} = \begin{pmatrix} ba_{11} & ba_{12} \\ ba_{21} & ba_{22} \end{pmatrix}$$

General matrix multiplication

Two matrices can be multiplied together provided the number of columns in the first matrix is equal to the number of rows in the second matrix.

$$\begin{pmatrix} a_{11} & a_{12} & a_{13} \\ a_{21} & a_{22} & a_{23} \end{pmatrix} \begin{pmatrix} b_{11} & b_{12} \\ b_{21} & b_{22} \\ b_{31} & b_{32} \end{pmatrix}$$

$$= \begin{pmatrix} a_{11}b_{11} + a_{12}b_{21} + a_{13}b_{31} & a_{11}b_{12} + a_{12}b_{22} + a_{13}b_{32} \\ a_{21}b_{11} + a_{22}b_{21} + a_{23}b_{31} & a_{21}b_{12} + a_{22}b_{22} + a_{23}b_{32} \end{pmatrix}$$

If matrix **A** is of order $(p \times q)$ and matrix **B** is of order $(q \times r)$ then if **C** = **AB**, the order of **C** is $(p \times r)$.

Transposition of a matrix

When the rows of a matrix are interchanged with its columns the matrix is said to be 'transposed'. If the original matrix is denoted by **A**, its transpose is denoted by **A'** or $\mathbf{A}^T$.

$$\text{If } \mathbf{A} = \begin{pmatrix} a_{11} & a_{12} & a_{13} \\ a_{21} & a_{22} & a_{23} \end{pmatrix} \text{ then } \mathbf{A}^T = \begin{pmatrix} a_{11} & a_{21} \\ a_{12} & a_{22} \\ a_{13} & a_{23} \end{pmatrix}$$

Adjoint of a matrix

If $\mathbf{A} = [a_{ij}]$ is any matrix and $\mathbf{A}_{ij}$ is the cofactor of a_{ij}, the matrix $[\mathbf{A}_{ij}]^T$ is called the *adjoint* of **A**. Thus

$$\mathbf{A} = \begin{pmatrix} a_{11} & a_{12} & \dots & a_{1n} \\ a_{21} & a_{22} & \dots & a_{2n} \\ . & . & & . \\ . & . & & . \\ . & . & & . \\ a_{n1} & a_{n2} & \dots & a_{nn} \end{pmatrix} \quad \text{adj } \mathbf{A} = \begin{pmatrix} A_{11} & A_{21} & \dots & A_{n1} \\ A_{12} & A_{22} & \dots & A_{n2} \\ . & . & & . \\ . & . & & . \\ . & . & & . \\ A_{1n} & A_{2n} & \dots & A_{nn} \end{pmatrix}$$

Singular matrix

A square matrix is singular if the determinant of its coefficients is zero.

The inverse of a matrix

If $\mathbf{A}$ is a non-singular matrix of order $(n \times n)$ then its inverse is denoted by $\mathbf{A}^{-1}$ such that

$$\mathbf{A}\mathbf{A}^{-1} = \mathbf{I} = \mathbf{A}^{-1}\mathbf{A}$$

$$\mathbf{A}^{-1} = \frac{\text{adj } (\mathbf{A})}{\Delta} \qquad \Delta = \det\ (\mathbf{A})$$

$$\mathbf{A}_{ij} = \text{ cofactor of } \mathrm{a}_{ij}$$

$$\text{If } \mathbf{A} = \begin{pmatrix} a_{11} & a_{12} & \dots & a_{1n} \\ a_{21} & a_{22} & \dots & a_{2n} \\ . & . & \dots & . \\ . & . & \dots & . \\ . & . & \dots & . \\ a_{n1} & a_{n2} & \dots & a_{nn} \end{pmatrix} \quad \mathbf{A}^{-1} = \frac{1}{\Delta} \begin{pmatrix} A_{11} & A_{21} & \dots & A_{n1} \\ A_{12} & A_{22} & \dots & A_{n2} \\ . & . & \dots & . \\ . & . & \dots & . \\ . & . & \dots & . \\ A_{1n} & A_{2n} & \dots & A_{nn} \end{pmatrix}$$

Solutions of simultaneous linear equations

The set of linear equations

$$\begin{matrix} a_{11}\boldsymbol{x}_1 + a_{12}\boldsymbol{x}_2 + \dots + a_{1n}\boldsymbol{x}_n = \boldsymbol{b}_1 \\ a_{21}\boldsymbol{x}_1 + a_{22}\boldsymbol{x}_2 + \dots + a_{2n}\boldsymbol{x}_n = \boldsymbol{b}_2 \\ \vdots \qquad \vdots \qquad\qquad \vdots \quad \vdots \\ a_{n1}\boldsymbol{x}_1 + a_{n2}\boldsymbol{x}_2 + \dots + a_{nn}\boldsymbol{x}_n = \boldsymbol{b}_n \end{matrix}$$

where the as and $\boldsymbol{b}$s are known, may be represented by the single matrix equation $\mathbf{A}x = \boldsymbol{b}$, where $\mathbf{A}$ is the $(n \times n)$ matrix of coefficients, a_{ij}, and $\boldsymbol{x}$ and $\boldsymbol{b}$ are $(n \times 1)$ column vectors. The solution to this matrix equation, if $\mathbf{A}$ is non-singular, may be written as $\boldsymbol{x} = \mathbf{A}^{-1}\boldsymbol{b}$ which leads to a solution given by *Cramer's rule*

$$x_i = \det \mathbf{D}_i / \det \mathbf{A}_i = 1, 2, ..., n$$

where det $\mathbf{D}_i$ is the determinant obtained from det **A** by replacing the elements of a_{ki} of the *i*th column by the elements b_k ($k = 1, 2, \ldots, n$). Note that this rule is obtained by using $\mathbf{A}^{-1} = (\det \mathbf{A})^{-1}$ adj **A** and so again is of practical use only when $n \le 4$.

If det **A** = 0 but det $\mathbf{D}_i \neq 0$ for some *i* then the equations are inconsistent: for example

$x + y = 2, x + y = 3$ has no solution.

Ordinary differential equations

A differential equation is a relation between a function and its derivatives. The order of the highest derivative appearing is the order of the differential equation. Equations involving only one independent variable are 'ordinary' differential equations, whereas those involving more than one are 'partial' differential equations.

If the equation involves no products of the function with its derivatives or itself nor of derivatives with each other, then it is 'linear'. Otherwise it is 'non-linear'.

A linear differential equation of order *n* has the form

$$P_0 \frac{\mathrm{d}^n y}{\mathrm{d}x^n} + P_1 \frac{\mathrm{d}^{n-1} y}{\mathrm{d}x^{n-1}} + \cdots + P_{n-1} \frac{\mathrm{d}y}{\mathrm{d}x} + P_n y = F$$

where $P_i (i = 0, 1, ..., n)$, F may be functions of x or constants, and $P_0 \neq 0$.

First order differential equations

Form	Type	Method
$\frac{dy}{dx} = f\left(\frac{y}{x}\right)$	Homogeneous	Substitute $u = \frac{y}{x}$
$\frac{dy}{dx} = f(x)g(y)$	Separable	$\int \frac{dy}{g(y)} = \int f(x)dx + C$ note that roots of $g(y) = 0$ are also solutions

$g(x,y)\frac{dy}{dx}+f(x,y)=0$ and $\frac{\partial f}{\partial y}=\frac{\partial g}{\partial x}$	Exact	Put $\frac{\partial \phi}{\partial x}=f$ and $\frac{\partial \phi}{\partial y}=g$ and solve these equations for ϕ. $\phi(x, y)$ = constant is the solution
$\frac{dy}{dx}+f(x)y=g(x)$	Linear	Multiply through by $p(x)=\exp(\int^x f(t)dt)$ giving $p(x)y=\int^x g(s)p(s)ds+C$

Second order (linear) equations

These are of the form

$$P_0(x)\frac{d^2y}{dx^2}+P_1(x)\frac{dy}{dx}+P_2(x)y=F(x)$$

When P_0, P_1, P_2 are constants and $f(x) = 0$, the solution is found from the roots of the auxiliary equation

$$P_0 m_2 + P_1 m + P_2 = 0$$

There are three other cases:

(i) Roots $m = \alpha$ and β are real and $\alpha \neq \beta$
$y(x) = Ae^{\alpha x} + Be^{\beta x}$

(ii) Double roots: $\alpha = \beta$
$y(x) = (A + Bx)\, e^{ax}$

(iii) Roots are complex: $m = k \pm il$
$y(x) = (A \cos lx + B \sin lx)e^{kx}$

Laplace transforms

If $f(t)$ is defined for all t in $0 \leq t < \infty$, then

$$L[f(t)] = F(s) = \int_0^\infty e^{-st} f(t)dt$$

is called the *Laplace transform* of $f(t)$. The two functions of $f(t)$, $F(s)$ are known as a transform pair, and

$$f(t) = L^{-1}[F(s)]$$

is called the *inverse transform* of $F(s)$.

Function	Transform
$f(t)$, $g(t)$	$F(s)$, $G(s)$
$c_1 f(t) + c_2 g(t)$	$c_1 F(s) + c_2 G(s)$
$\int_0^t f(x)dx$	$F(s)/s$
$(-t)^n f(t)$	$\dfrac{d^n F}{ds^n}$
$e^{at} f(t)$	$F(s-a)$
$f(t-a)H(t-a)$	$e^{-as}F(s)$
$\dfrac{d^n f}{dt^n}$	$s^n F(s) - \sum_{r=1}^{n} s^{n-r} f^{(r-1)}(0+)$
$\dfrac{1}{a} e^{-bt} \sin at,\ a > 0$	$\dfrac{1}{(s+b)^2 + a^2}$
$e^{-bt} \cos at$	$\dfrac{s+b}{(s+b)^2 + a^2}$
$\dfrac{1}{a} e^{-bt} \sinh at,\ a > 0$	$\dfrac{1}{(s+b)^2 + a^2}$
$e^{-bt} \cosh at$	$\dfrac{s+b}{(s+b)^2 + a^2}$
$(\pi t)^{-1/2}$	$s^{-1/2}$
$\dfrac{2^n t^{n-1/2}}{1 \cdot 3 \cdot 5 \ldots (2n-1)\sqrt{\pi}}$, n integer	$s^{-(n+1/2)}$
$\dfrac{\exp(-a^2/4t)}{2(\pi t^3)^{1/2}}\ (a > 0)$	$e^{-a\sqrt{s}}$

Basic trigonometry

Definitions (See Fig. 1.9)

sine: $\sin A = \dfrac{y}{r}$ cosine: $\cos A = \dfrac{x}{r}$

tangent: $\tan A = \dfrac{y}{x}$ cotangent: $\cot A = \dfrac{x}{y}$

secant: $\sec A = \dfrac{r}{x}$ cosecant: $\operatorname{cosec} A = \dfrac{r}{y}$

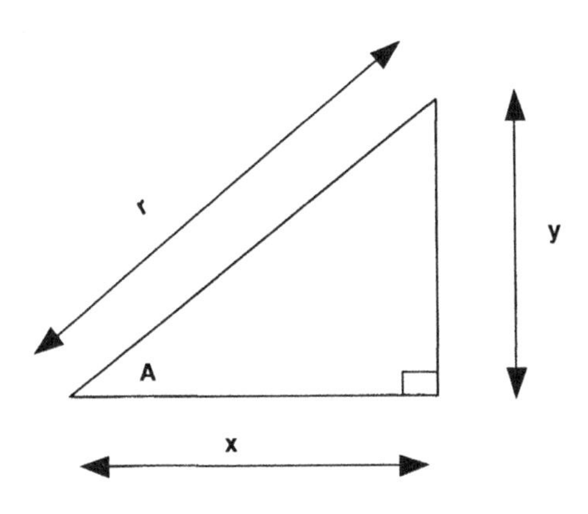

Fig. 1.9 Basic trigonometry

Relations between trigonometric functions

$\sin^2 A + \cos^2 A = 1 \quad \sec^2 A = 1 + \tan^2 A$

$\operatorname{cosec}^2 A = 1 + \cot^2 A$

	$\sin A = s$	$\cos A = c$	$\tan A = t$
$\sin A$	s	$(1 - c^2)^{1/2}$	$t(1 + t^2)^{-1/2}$
$\cos A$	$(1 - s^2)^{-1/2}$	c	$(1 + t^2)^{-1/2}$
$\tan A$	$s(1 - s^2)^{-1/2}$	$(1 - c^2)^{1/2}/c$	t

A is assumed to be in the first quadrant; signs of square roots must be chosen appropriately in other quadrants.

Addition formulae

$\sin(A \pm B) = \sin A \cos B \pm \cos A \sin B$
$\cos(A \pm B) = \cos \mathrm{A} \cos B \mp \sin A \sin B$

$$\tan(A \pm B) = \frac{\tan A \pm \tan B}{1 \mp \tan A \tan B}$$

Sum and difference formulae

$\sin A + \sin B = 2\sin \tfrac{1}{2}(A + B) \cos \tfrac{1}{2}(A - B)$
$\sin A - \sin B = 2\cos \tfrac{1}{2}(A + B) \sin \tfrac{1}{2}(A - B)$
$\cos A + \cos B = 2\cos \tfrac{1}{2}(A + B) \cos \tfrac{1}{2}(A - B)$
$\cos A - \cos B = 2\sin \tfrac{1}{2}(A + B) \sin \tfrac{1}{2}(B - A)$

Product formulae

$\sin A \sin B = \tfrac{1}{2}\{\cos(A - B) - \cos(A + B)\}$
$\cos A \cos B = \tfrac{1}{2}\{\cos(A - B) + \cos(A + B)\}$
$\sin A \cos B = \tfrac{1}{2}\{\sin(A - B) + \sin(A + B)\}$

Powers of trigonometric functions

$\sin^2 A = \tfrac{1}{2} - \tfrac{1}{2}\cos 2A$
$\cos^2 A = \tfrac{1}{2} + \tfrac{1}{2}\cos 2A$
$\sin^3 A = \tfrac{3}{4}\sin A - \tfrac{1}{4}\sin 3A$
$\cos^3 A = \tfrac{3}{4}\cos A + \tfrac{1}{4}\cos 3A$

Co-ordinate geometry

Straight-line

General equation	$ac + by + c = 0$	m = gradient c = intercept on the y-axis
Gradient equation	$y = mx + c$	
Intercept equation	$\frac{x}{A} + \frac{y}{B} = 1$	A = intercept on the x-axis B = intercept on the y-axis
Perpendicular equation	$x \cos \alpha + y \sin \alpha = p$	p = length of perpendicular from the origin to the line α = angle that the perpendicular makes with the x-axis

The distance between two points $P(x_1, y_1)$ and $Q(x_2, y_2)$ is given by

$$PQ = \sqrt{[(x_1 - x_2)^2 + (y_1 - y_2)^2]}$$

The equation of the line joining two points (x_1, y_1) and (x_2, y_2) is given by

$$\frac{y - y_1}{y_1 - y_2} = \frac{x - x_1}{x_1 - x_2}$$

Circle

General equation $x^2 + y^2 + 2gx + 2fy + c = 0$
The centre has co-ordinates $(-g, -f)$
The radius is $r = \sqrt{(g^2 + f^2 - c)}$
The equation of the tangent at (x_1, y_1) to the circle is

$$xx_1 + yy_1 + g(x + x_1) + f(y + y_1) + c = 0$$

The length of the tangent from (x_1, y_1) to the circle is

$$t^2 = x_1^2 + y_1^2 + 2gx_1 + 2fy_1 + c$$

Parabola (see Fig. 1.10)

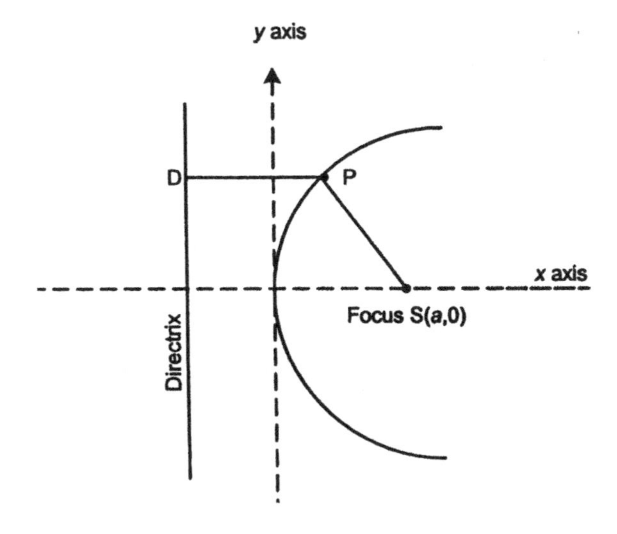

Fig. 1.10 Parabola

Eccentricity $e = \frac{SP}{PD} = 1$

With focus $S(a, 0)$ the equation of a parabola is $y^2 = 4ax$

The parametric form of the equation is $x = at^2$, $y = 2at$

The equation of the tangent at (x_1, y_1) is $yy_1 = 2a(x + x_1)$

Ellipse (see Fig. 1.11)

Eccentricity $e = \frac{SP}{PD} < 1$

The equation of an ellipse is $\frac{x^2}{a^2} + \frac{y^2}{b^2} = 1$ where $b^2 = a^2(1 - e^2)$

The equation of the tangent at (x_1, y_1) is $\frac{xx_1}{a^2} + \frac{yy_1}{b^2} = 1$

The parametric form of the equation of an ellipse is $x = a \cos \theta$, $y = b \sin \theta$, where θ is the eccentric angle

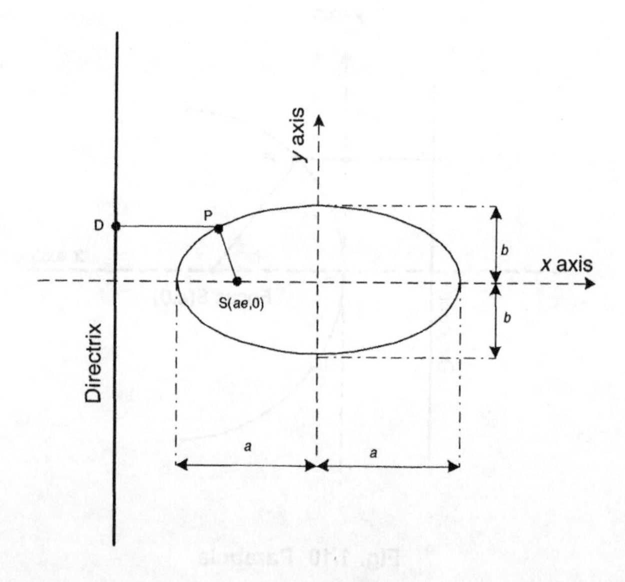

Fig. 1.11 Ellipse

Hyperbola (see Fig. 1.12)

Eccentricity $e = \dfrac{SP}{PD} > 1$

The equation of a hyperbola is $\dfrac{x^2}{a^2} - \dfrac{y^2}{b^2} = 1$ where $b^2 = a^2(e^2 - 1)$

The parametric form of the equation is $x = a \sec \theta$, $y = b \tan \theta$, where θ is the eccentric angle

The equation of the tangent at (x_1, y_1) is $\dfrac{xx_1}{a^2} - \dfrac{yy_1}{b^2} = 1$

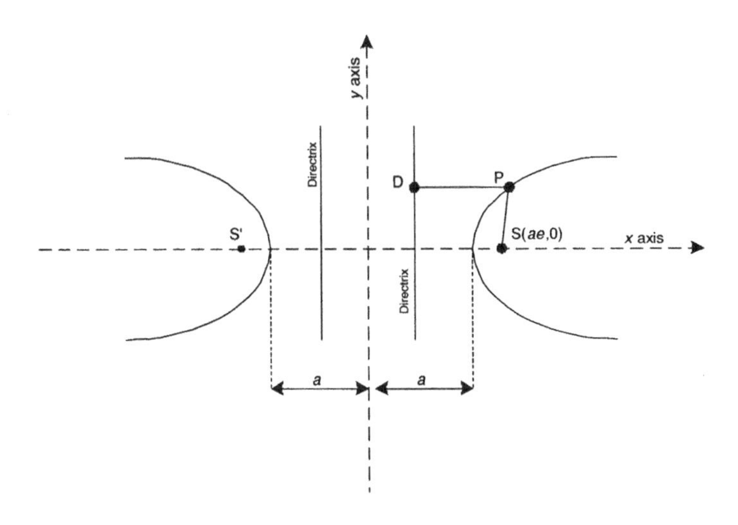

Fig. 1.12 Hyperbola

Sine wave (see Fig. 1.13)

$$y = a \sin (bx + c)$$

$$y = a \cos (bx + c') = a \sin (bx + c) \quad (\text{where } c = c' + \pi/2)$$

$$y = m \sin bx + n \cos bx = a \sin (bx + c)$$

$$[\text{where } a = \sqrt{(m^2 + n^2)},\ c = \tan^{-1}(n/m)]$$

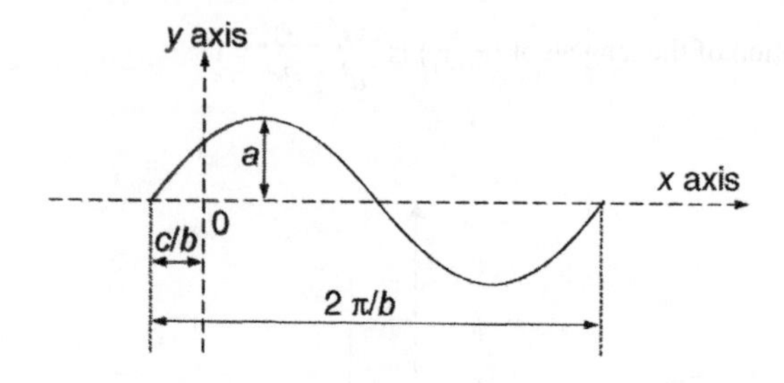

Fig. 1.13 Sine wave

Helix (see Fig. 1.14)

A helix is a curve generated by a point moving on a cylinder with the distance it transverses parallel to the axis of the cylinder being proportional to the angle of rotation about the axis

$$x = a \cos \theta$$

$$y = a \sin \theta$$

$$z = k\theta$$

(where a = radius of cylinder, $2\pi k$ = pitch)

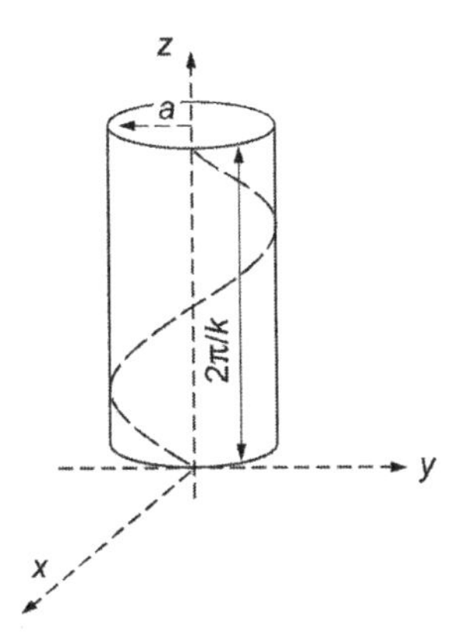

Fig. 1.14 Helix

1.9 Useful references and standards

For links to *The Reference Desk*, a website containing over 6000 on-line units conversions 'calculators', go to:

www.flinthills.com/~ramsdale/EngZone/refer.htm

Standards

1. ASTM/IEEE SI 10: 1997 *Use of the SI system of units* (replaces ASTM E380 and IEEE 268).

CHAPTER 2

Bending, Torsion, and Stress

2.1 Simple stress and strain

Stress, $\sigma = \frac{\text{load}}{\text{area}} = \frac{P}{A}$ units are N/m^2 (see Fig. 2.1)

Strain, $\varepsilon = \frac{\text{change in length}}{\text{original length}} = \frac{d}{l}$ = a ratio, therefore, no units

Hooke's law: $\frac{\text{stress}}{\text{deformation}}$ = constant

= Young's modulus E N/m^2

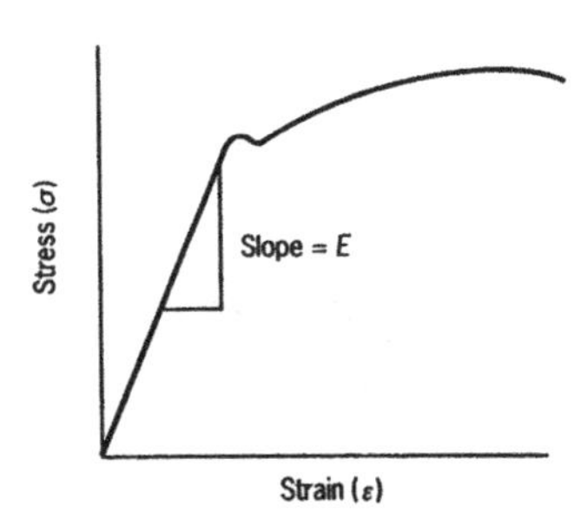

Fig. 2.1 Stress and strain

Poisson's ratio (ν) = $\frac{\text{lateral strain}}{\text{longitudinal strain}} = \frac{\delta d/d}{\delta l/l}$

(ratio, therefore, no units; see Fig. 2.2)

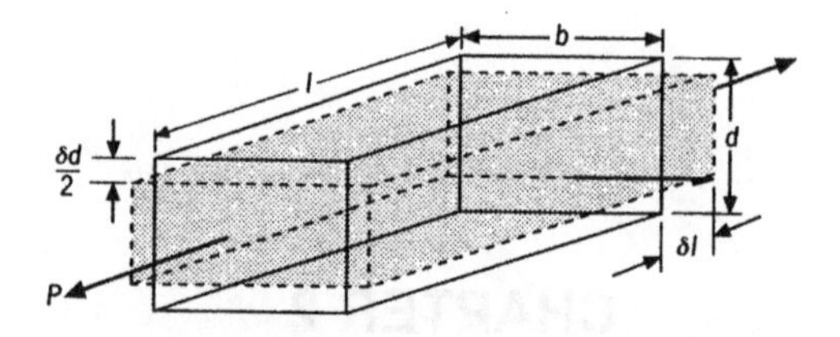

Fig. 2.2 Poisson's ratio

Shear stress (τ) = $\dfrac{\text{shear load}}{\text{area}} = \dfrac{Q}{A}$: units: N/m^2 (see Fig. 2.3)

Shear strain (γ) = angle of deformation under shear stress

Modulus of rigidity= $\dfrac{\text{shear stress}}{\text{shear strain}} = \dfrac{\tau}{\gamma}$

= constant G (units are N/m^2)

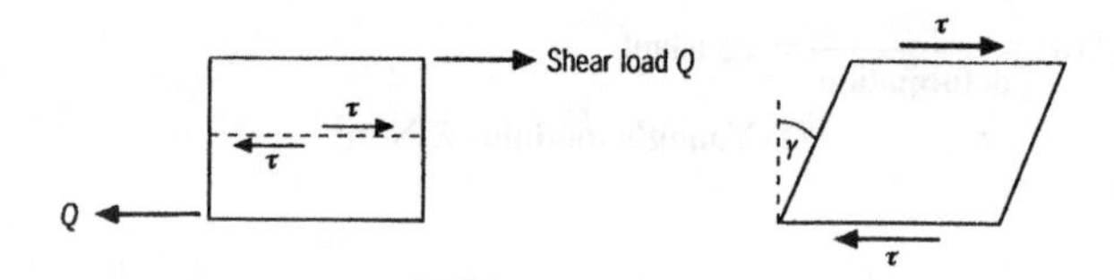

Fig. 2.3 Shear stress

2.2 Simple elastic bending (flexure)

The simple theory of elastic bending is

$$\frac{M}{I} = \frac{\sigma}{y} = \frac{E}{R}$$

M = applied bending moment
I = second moment about the neutral axis
R = radius of curvature of neutral axis
E = Young's modulus
σ = stress due to bending at distance y from neutral axis

The second moment of area is defined, for any section, as

$$I = y^2 \mathrm{d}A$$

I for common sections is calculated as follows in Fig. 2.4.

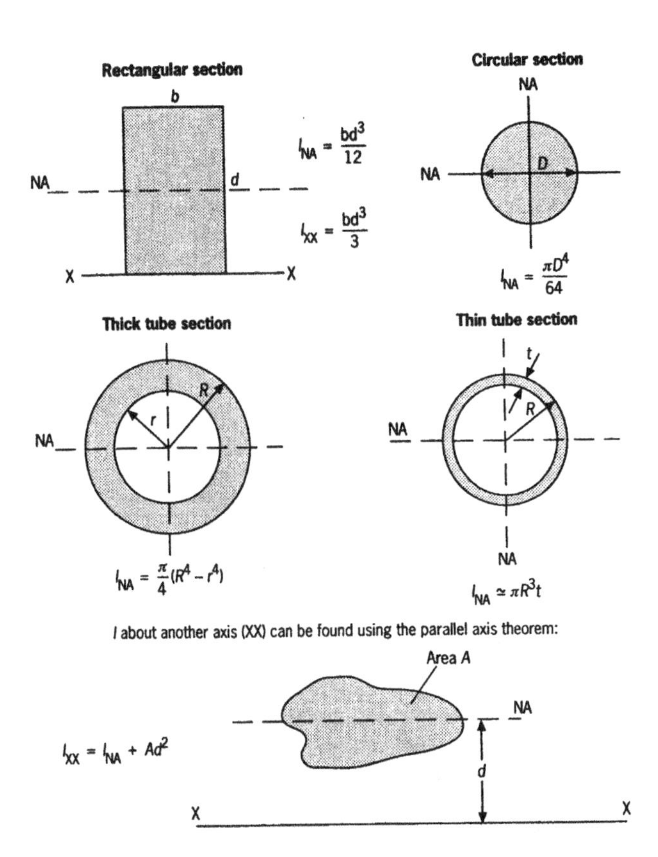

Fig. 2.4 *I* for common sections

Deformable Body Mechanics

Steelwork sections

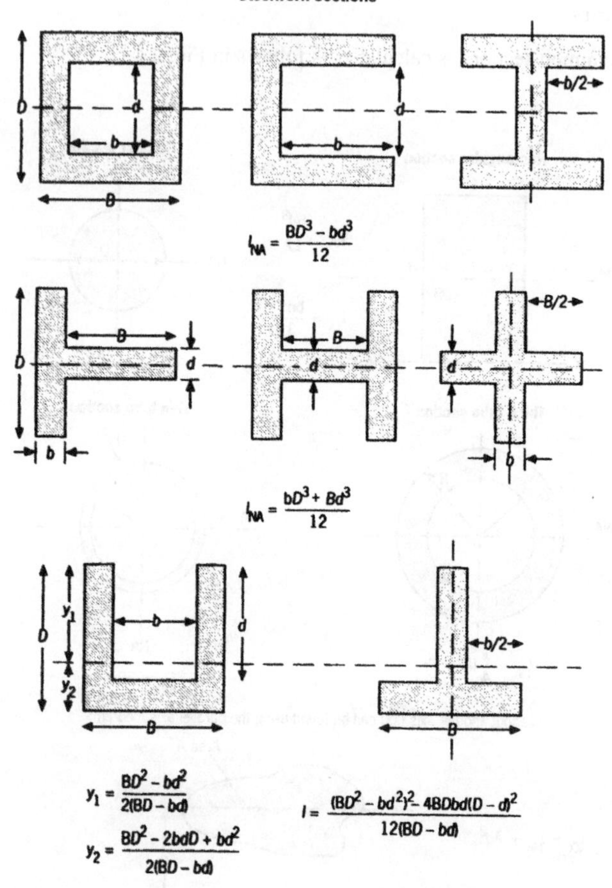

Fig. 2.4 ***I*** **for common sections (cont.)**

Section modulus Z is defined as

$$Z = \frac{I}{y}$$

Strain energy due to bending, U, is defined as

$$U = \int_0^1 \frac{M^2 \mathrm{d}s}{2EI}$$

For uniform beams subject to constant bending moment this reduces to

$$U = \frac{M^2 l}{2EI}$$

2.3 Slope and deflection of beams

Many rotating equipment components (shafts, blades, bearings, etc.) can be modelled as simple beams.

The relationships between load W, shear force SF, bending moment M, slope, and deflection are:

$$\text{Deflection} = \delta \text{ (or } y\text{)}$$

$$\text{Slope} = \frac{\mathrm{d}y}{\mathrm{d}x}$$

$$M = EI \frac{\mathrm{d}^2 y}{\mathrm{d}x^2}$$

$$F = EI \frac{\mathrm{d}^3 y}{\mathrm{d}x^3}$$

$$W = EI \frac{\mathrm{d}^4 y}{\mathrm{d}x^4}$$

Values for common beam configurations are shown in Fig. 2.5.

2.4 Torsion

A torsional moment is a common occurrence in rotating equipment design and can be treated in much the same way as bending, i.e. torsional moment or torque (T) can be equated to a stress gradient multiplied by a second moment of area. In this case the second moment (J) lies in the plane of stress and is called the 'polar second moment of area' or 'polar moment of inertia' of the section. The stress in these conditions is shear stress, whose sign (i.e. rotational tendency) reverses from one side of the centroid to the other.

Conditions of support and loading	Bending moment (maximum)	Shearing force (maximum)	Safe load W	Deflection (maximum)
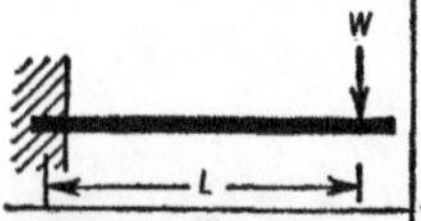 W; L	WL	W	$\frac{M}{L}$	$\frac{WL^3}{3EI}$
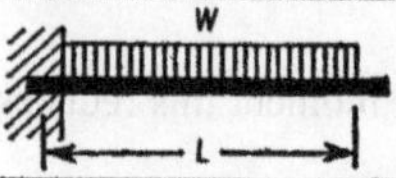 W; L	$\frac{WL}{2}$	W	$\frac{2M}{L}$	$\frac{WL^3}{8EI}$
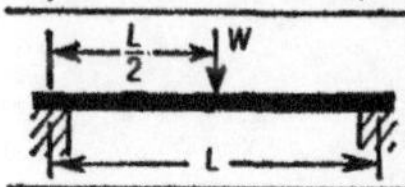 $\frac{L}{2}$; W; L	$\frac{WL}{4}$	$\frac{W}{2}$	$\frac{4M}{L}$	$\frac{WL^3}{48EI}$
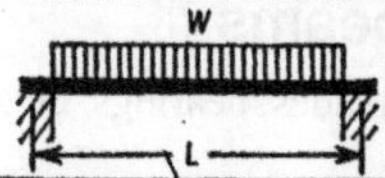 W; L	$\frac{WL}{8}$	$\frac{W}{2}$	$\frac{8M}{L}$	$\frac{5WL^3}{384EI}$
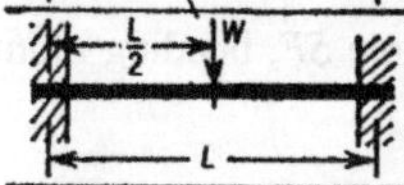 $\frac{L}{2}$; W; L	$\frac{WL}{8}$	$\frac{W}{2}$	$\frac{8M}{L}$	$\frac{WL^3}{192EI}$
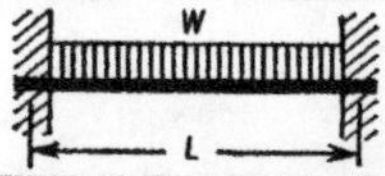 W; L	$\frac{WL}{12}$	$\frac{W}{2}$	$\frac{12M}{L}$	$\frac{WL^3}{384EI}$
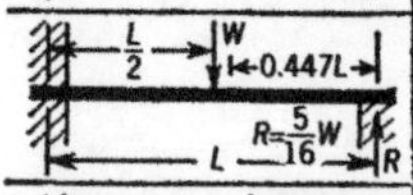 $\frac{L}{2}$; W; 0.447L; $R=\frac{5}{16}W$; L; R	$\frac{3WL}{16}$	$\frac{11W}{16}$	$\frac{16M}{3L}$	$\frac{WL^3}{107EI}$
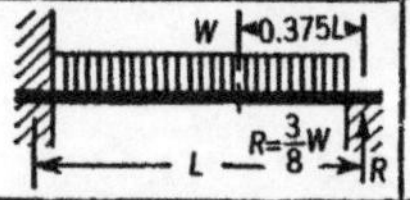 W; 0.375L; $R=\frac{3}{8}W$; L; R	$\frac{WL}{8}$	$\frac{5W}{8}$	$\frac{8M}{L}$	$\frac{WL^3}{187EI}$

Fig. 2.5 Slope and deflection of beams

For solid or hollow shafts of uniform cross-section, the torsion formula is (see Figs 2.6 and 2.7)

$$\frac{T}{J} = \frac{\tau}{R} = \frac{G\theta}{l}$$

T = torque applied (Nm)
J = polar second moment of area (m^4)
τ = shear stress (N/m^2)
R = radius (m)
G = modulus of rigidity (N/m^2)
θ = angle of twist (rad)
l = length (m)

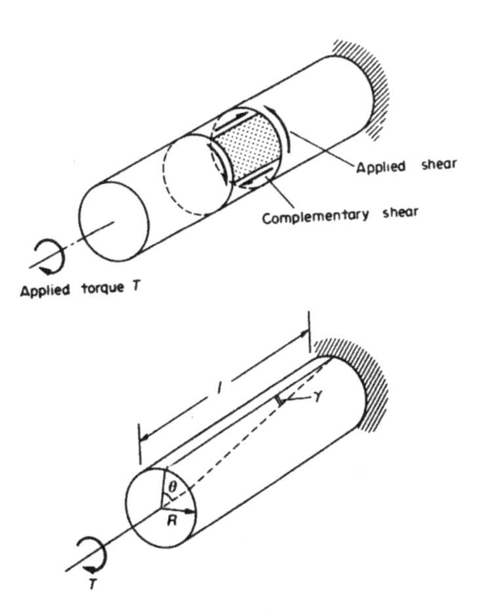

Fig. 2.6 Torsion

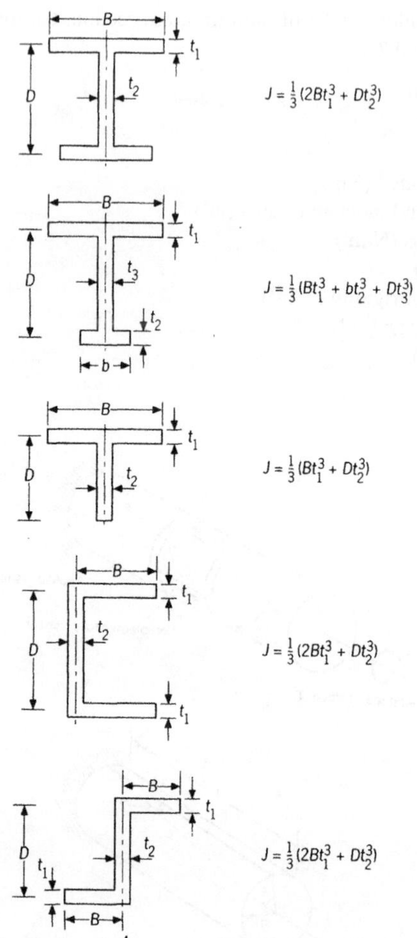

The polar second motion of area $(J)m^4$ is a measure of the stiffness of a member in pure twisting

Fig. 2.7 Torsion formulae

For solid shafts

$$J = \frac{\pi D^4}{32}$$

For hollow shafts

$$J = \frac{\pi \left(D^4 - d^4\right)}{32}$$

For thin-walled hollow shafts

$$J \cong 2\pi r^3 t$$

where
r = mean radius of shaft wall
t = wall thickness

Solid shaft with flange

Figure 2.8 shows the situation of a flanged shaft subjected to concentric torsional loading. This is typically analysed as a system where the flange is regarded as being rigidly held at its face and the flange thickness designed so the stress at the junction of the shaft and flange is broadly equal to that at the surface of the shaft. Hence, the design criteria become:

- area of possible fracture = πdt
- shear resistance = $\pi dt \tau_{max}$
- shear resistance moment (T_f) = $\pi dt \tau_{max} \times d/2$
- for a balanced design, torque in flange (T_f) = torque in shaft (T_r) hence using

$$\frac{T}{J} = \frac{\tau}{R}$$

and

$$J = \frac{\pi d^4}{32}$$

$$T_r = \frac{\pi d^3}{16} \tau_{max}$$

Then

$$\frac{\pi}{2} d^2 t \tau_{max} \cong \frac{\pi d^3}{16} \tau_{max}$$

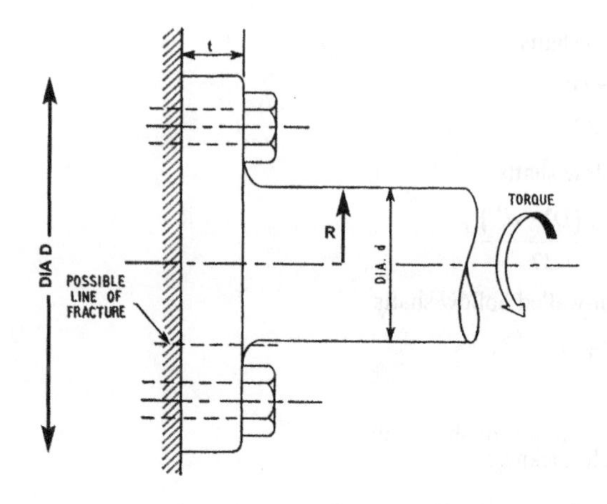

Fig. 2.8 Torsional loading of a flanged shaft

Hence as a 'rule of thumb' for flange design; theoretical flange thickness $\geq d/8$. In practice (and in flange design codes), the flange thickness is increased above this minimum value to allow for the weakening effect of flange bolt holes and the need for significant flange-to-flange bolting forces.

Strain energy (U) in torsion

In torsion, strain energy (U) is expressed as

$$U = \frac{T^2 l}{2GJ} = \frac{GJ\theta^2}{2l}$$

Torsion of non-circular sections

If a section concentrically loaded in torsion is not of uniform circular shape, the stress distribution is not a simple case. Since projections cannot carry any stress at their tips, a stress gradient must exist between each tip and the adjacent points of maximum stress. Stress is generally assumed to become a maximum at approximately the greatest distance from the centre at which a continuous circular annulus can be formed within the section. The stress then varies uniformly between its maximum value and zero at the axis, and between its maximum value and zero at the projection extremities. The actual variation depends on the geometry of the projections. Torsion of a plane rectangular section provides a good illustration.

Torsion of rectangular sections

Figure 2.9 shows a rectangular section of dimensions $a \times b$. The maximum shear stress (τ_{max}) occurs at the middle of the long sides at point X, where the largest continuous annulus can form. There is another smaller maximum at the middle of the short sides at point Y. Stress at the corners and at the centre is zero and the stress distribution over the sides is *approximately* parabolic, as shown.

The relationship between the twisting moment T and the maximum shear stress τ_{max} at X is approximately

$$T = \mathrm{K}ab^2\tau_{max}$$

where constant $\mathrm{K} \cong \dfrac{1}{3+1.8\dfrac{b}{a}}$

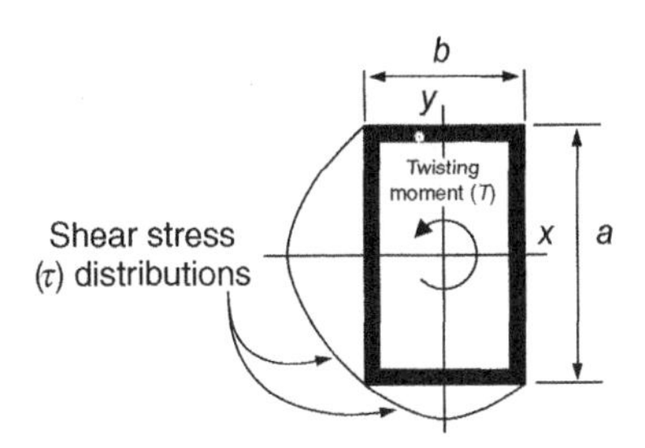

Fig. 2.9 Stress distribution over a rectangular section under torsion

Transmission of torque using keyed couplings

In a keyed coupling, the transmission of torque between concentric components is achieved by means of transverse shear stress acting through a longitudinal pin or key as shown in Fig. 2.10. If a single key is likely to lead to unbalance in loading or weakness in the shaft section, two diametrically opposed square keys of smaller size may be used. In certain applications the key may be integral with the shaft, i.e. where stress intensity is exceptionally high or the components need to slide relative to each other. In this case the key is called a 'spline'. The logical conclusion of this concept is a splined shaft having a series of uniformly circumferentially spaced splines engaging with a corresponding female socket, as shown in Fig. 2.11. Some broad 'rules of thumb' in key sizing are:

1. The maximum key width w may be taken as $d/4$ where d is the shaft diameter.
2. The effective key length (l) may be approximated to $3d/2$ and the width adjusted accordingly. With this proportion $w \leq d/4$.

In either case the calculated key length should be increased to allow for the rounded ends of the key (i.e. it will not provide full drive over its full length).

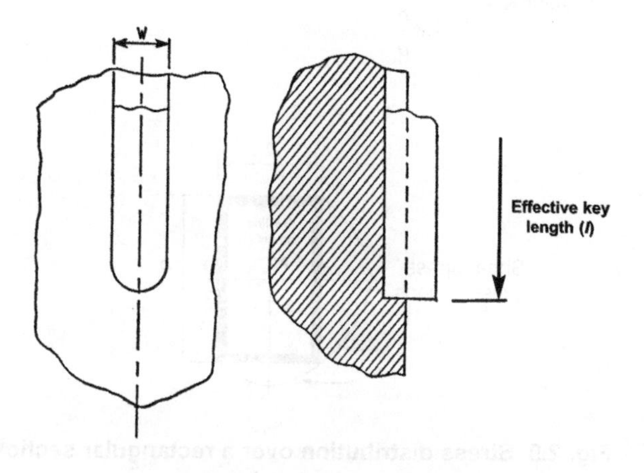

Fig. 2.10 A square key end shape

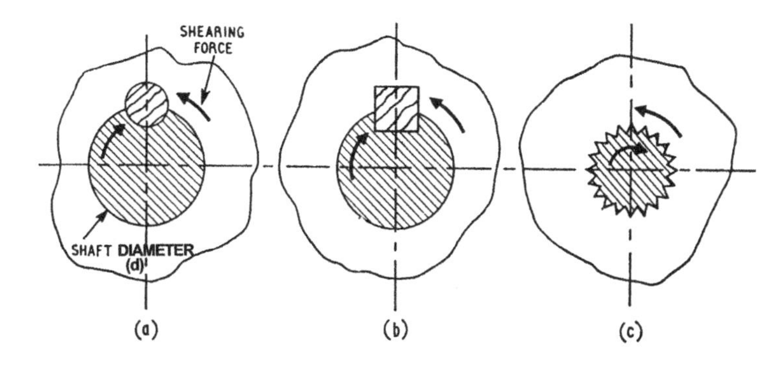

Fig. 2.11 'Keyed' drives on shafts (a) A circular key (b) A square key (c) A splined shaft

Single-key, loose-flange couplings

Figure 2.12 shows typical outline design dimensions for a single-keyed loose-flange coupling. Some typical design equations and crude rules of thumb are:

- boss diameter (D) > $2d$ to allow for the keyway
- torque resistance of the key (T_k) $\cong \frac{ld^2}{8}\tau_k$ when τ_k = allowable shear stress in the key
- bolt shear resistance $\cong \frac{(n-1)\pi\delta^2\tau_{bolt}}{4}$
  where
  n = number of bolts
  δ = bolt diameter
- the pitch radius (R) of the bolts and the bolt diameter (δ) are found by simple trial and error in the above equation. Normally, for a solid shaft, a bolt diameter (δ) of $\delta = d/6$ is a good starting point. Another typical guideline calculation is

  $$\text{bolt diameter } (\delta) = \frac{0.423d}{\sqrt{n}} + 8 \text{ mm}$$

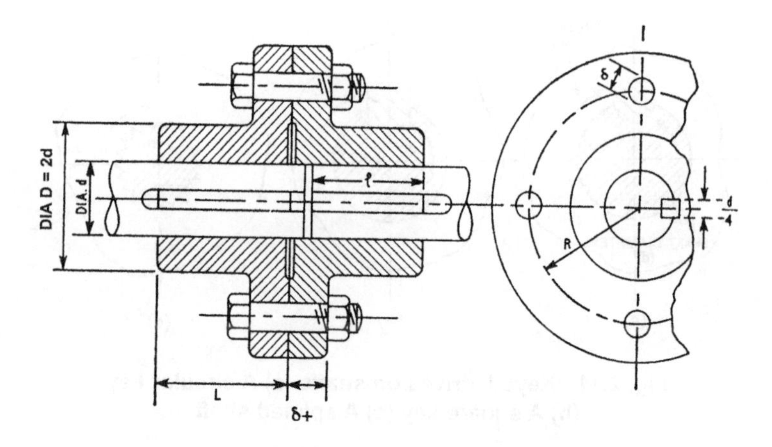

Fig. 2.12 Single-keyed loose-flange coupling: typical arrangement and dimensions

2.5 Combined bending and torsion

In most practical rotating equipment applications, the effects of bending and torsion do not exist in isolation, but are combined. The overall result is to increase stress (and resulting fatigue) loadings, thereby increasing the necessary factor of safety that has to be built in to the design if the equipment is to perform satisfactorily.

For a shaft of diameter, *d*, in combined bending and torsion the following equations are used:

Maximum resultant shear stress

$$\tau = \sqrt{\left(\frac{p^2}{4} + q^2\right)}$$

where

p = tensile or compressive stress

q = shear stress acting on the same plane as p

Maximum safe shear stress

$$\tau_{\max} = \frac{16T_{\mathrm{E}}}{\pi d^3}$$

where

$T_E = \sqrt{(M^2 + T^2)}$ termed the 'equivalent torque' resulting from bending and moment, M, and torque, T.

Figure 2.13 shows a typical application of equivalent torque, T_E, criterion for a diesel engine crankshaft – a classic example of a combined bending and torsion loading system. The figure also shows approximate design dimensions in terms of the main journal diameter, D.

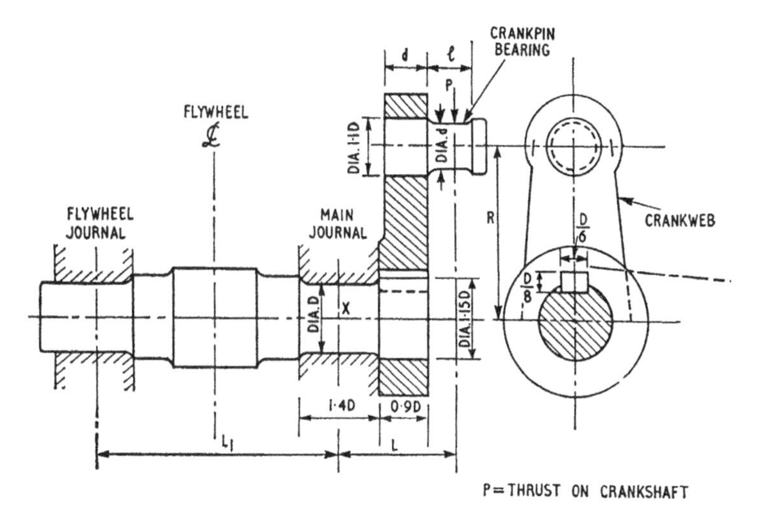

For the overhung crankshaft: Equivalent torque (T_E) $\cong P\sqrt{(L^2 + R^2)}$

Fig. 2.13 Crankshaft: some torque design 'rules of thumb'

2.6 Stress concentration factors

The effective stress in a component can be raised well above its expected levels owing to the existence of geometrical features causing stress concentrations under dynamic elastic conditions. Typical design stress concentration factors are as shown in Fig. 2.14.

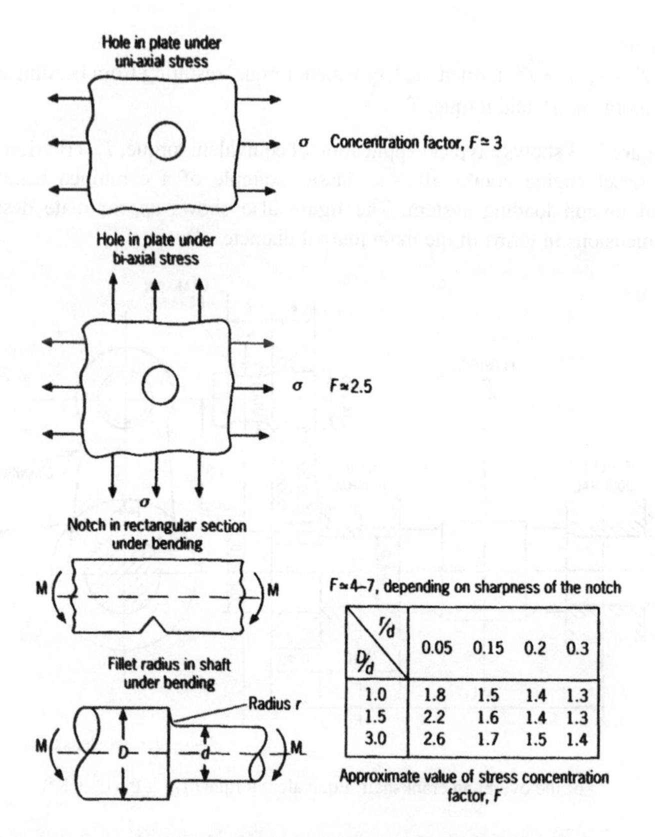

D/d \ r/d	0.05	0.15	0.2	0.3
1.0	1.8	1.5	1.4	1.3
1.5	2.2	1.6	1.4	1.3
3.0	2.6	1.7	1.5	1.4

Fig. 2.14 Stress concentration factors

APPROXIMATE STRESS CONCENTRATION FACTORS (Elastic Stresses)

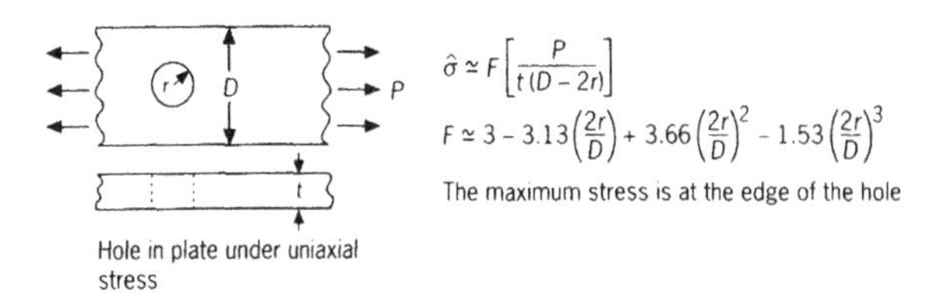

Hole in plate under uniaxial stress

$$\hat{\sigma} \simeq F\left[\frac{P}{t(D-2r)}\right]$$

$$F \simeq 3 - 3.13\left(\frac{2r}{D}\right) + 3.66\left(\frac{2r}{D}\right)^2 - 1.53\left(\frac{2r}{D}\right)^3$$

The maximum stress is at the edge of the hole

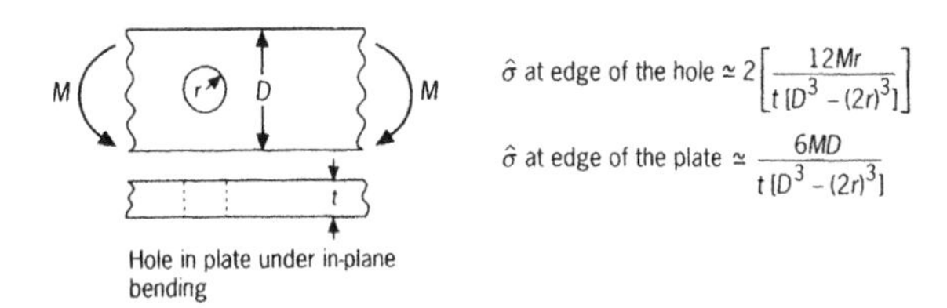

Hole in plate under in-plane bending

$$\hat{\sigma} \text{ at edge of the hole} \simeq 2\left[\frac{12Mr}{t[D^3-(2r)^3]}\right]$$

$$\hat{\sigma} \text{ at edge of the plate} \simeq \frac{6MD}{t[D^3-(2r)^3]}$$

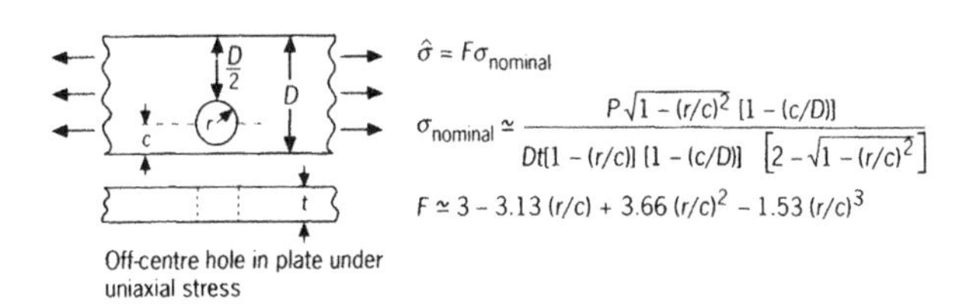

Off-centre hole in plate under uniaxial stress

$$\hat{\sigma} = F\sigma_{nominal}$$

$$\sigma_{nominal} \simeq \frac{P\sqrt{1-(r/c)^2}\,[1-(c/D)]}{Dt[1-(r/c)]\,[1-(c/D)]\,\left[2-\sqrt{1-(r/c)^2}\right]}$$

$$F \simeq 3 - 3.13\,(r/c) + 3.66\,(r/c)^2 - 1.53\,(r/c)^3$$

Fig. 2.14 Stress concentration factors (cont.)

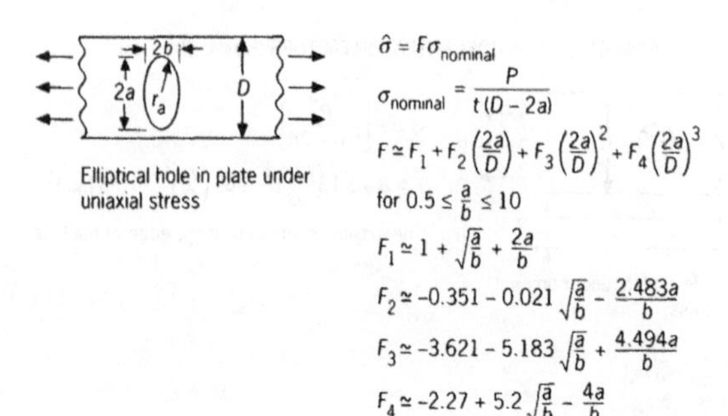

Elliptical hole in plate under uniaxial stress

$$\hat{\sigma} = F\sigma_{nominal}$$

$$\sigma_{nominal} = \frac{P}{t(D-2a)}$$

$$F \simeq F_1 + F_2\left(\frac{2a}{D}\right) + F_3\left(\frac{2a}{D}\right)^2 + F_4\left(\frac{2a}{D}\right)^3$$

for $0.5 \le \frac{a}{b} \le 10$

$$F_1 \simeq 1 + \sqrt{\frac{a}{b}} + \frac{2a}{b}$$

$$F_2 \simeq -0.351 - 0.021\sqrt{\frac{a}{b}} - \frac{2.483a}{b}$$

$$F_3 \simeq -3.621 - 5.183\sqrt{\frac{a}{b}} + \frac{4.494a}{b}$$

$$F_4 \simeq -2.27 + 5.2\sqrt{\frac{a}{b}} - \frac{4a}{b}$$

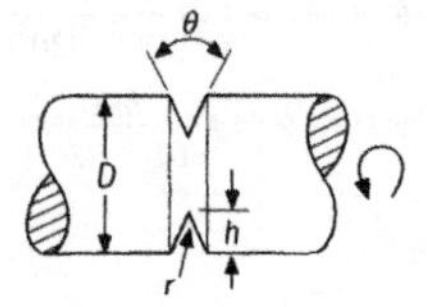

V-notch in circular shaft under torsion

F_v = stress concentration factor for V-notch

$$F_v \simeq F_u - \left[0.02 + 0.14\left(\frac{\theta}{135}\right)^2\right](F_u - 1)F_u$$

where F_u = stress concentration factor for U-notch in torsion

for $\frac{r}{D-2h} \le 0.01$ and $\theta \le 135^\circ$

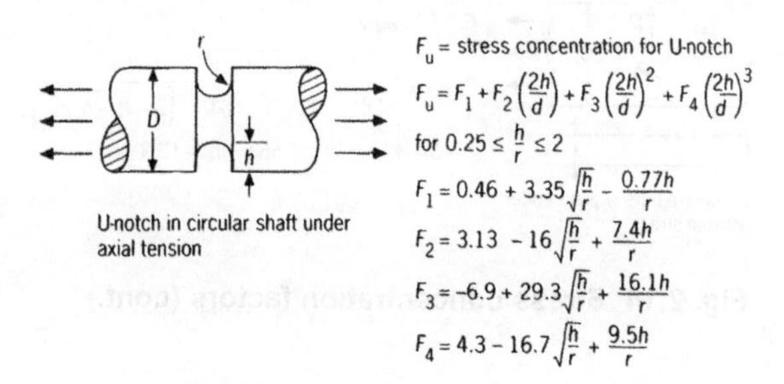

U-notch in circular shaft under axial tension

F_u = stress concentration for U-notch

$$F_u = F_1 + F_2\left(\frac{2h}{d}\right) + F_3\left(\frac{2h}{d}\right)^2 + F_4\left(\frac{2h}{d}\right)^3$$

for $0.25 \le \frac{h}{r} \le 2$

$$F_1 = 0.46 + 3.35\sqrt{\frac{h}{r}} - \frac{0.77h}{r}$$

$$F_2 = 3.13 - 16\sqrt{\frac{h}{r}} + \frac{7.4h}{r}$$

$$F_3 = -6.9 + 29.3\sqrt{\frac{h}{r}} + \frac{16.1h}{r}$$

$$F_4 = 4.3 - 16.7\sqrt{\frac{h}{r}} + \frac{9.5h}{r}$$

Fig. 2.14 Stress concentration factors (cont.)

CHAPTER 3

Motion and Dynamics

3.1 Making sense of dynamic equilibrium

The concept of dynamic equilibrium lies behind many types of engineering analyses and design of rotating equipment. Some key definition points are:

- Formally, an object is in a state of equilibrium when the forces acting on it are such as to leave it in its state of rest or uniform motion in a straight line.
- In terms of dynamic equilibrium, this means that it is moving at constant velocity with zero acceleration (or deceleration).

Figure 3.1 shows the difference between dynamic equilibrium and non-equilibrium. The concept of dynamic equilibrium is used to design individual components of rotating equipment.

3.2 Motion equations

Uniformly accelerated motion

Bodies under uniformally accelerated motion follow the general equations

$v = u + at$ $\quad t$ = time (s)

$s = ut + \frac{1}{2}at^2$ $\quad a$ = acceleration (m/s^2)

$s = \frac{u+v}{2}t$ $\quad s$ = distance travelled (m)

u = initial velocity (m/s)

$v^2 = u^2 + 2as$ $\quad v$ = final velocity (m/s)

Dynamic equilibrium

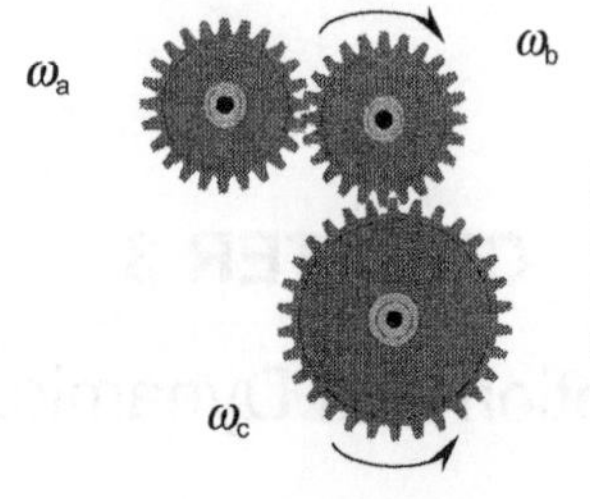

All parts of the mechanism are moving with constant angular velocities

Dynamic non-equilibrium

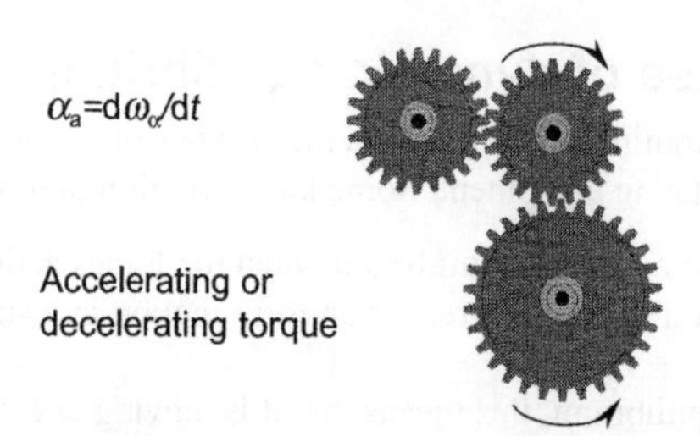

No parts of the mechanism are moving with constant velocity

Fig. 3.1 Dynamic equilibrium and non-equilibrium

Angular motion

$$\omega = \frac{2\pi N}{60}$$

$$\omega_2 = \omega_1 + \alpha$$

$$\theta = \frac{\omega_1 - \omega_2}{2} t$$

$$\omega_2^2 = \omega_1^2 + 2\alpha s$$

$$\theta = \omega_1 t + \tfrac{1}{2}\alpha^2$$

t = time (s)

θ = angle moved (rad)

α = angular acceleration (rad/s^2)

N = angular speed (rev/min)

ω_1 = initial angular velocity (rad/s)

ω_2 = final angular velocity (rad/s)

General motion of a particle in a plane

$v = \mathrm{d}s/\mathrm{d}t$	s = distance
$a = \mathrm{d}v/\mathrm{d}t = \mathrm{d}^2s/\mathrm{d}t^2$	t = time
$v = a\mathrm{d}t$	v = velocity
$s = v\mathrm{d}t$	a = acceleration

3.3 Newton's laws of motion

First law A body will remain at rest or continue in uniform motion in a straight line until acted upon by an external force.

Second law When an external force is applied to a body of constant mass it produces an acceleration that is directly proportional to the force, i.e. force (F) = mass (m) × acceleration (a).

Third law Every action produces an equal and opposite reaction.

Table 3.1 shows the comparisons between rotational and translational motion.

Table 3.1 Comparisons: rotational and translational motion

Translation		*Rotation*	
Linear displacement from a datum	x	Angular displacement	θ
Linear velocity	v	Angular velocity	ω
Linear acceleration	$a = \mathrm{d}v/\mathrm{d}t$	Angular acceleration	$\alpha = \mathrm{d}\omega/\mathrm{d}t$
Kinetic energy	$KE = mv^2/2$	Kinetic energy	$KE = I\omega^2/2$
Momentum	mv	Momentum	$I\omega$
Newton's second law	$F = m\mathrm{d}^2x/\mathrm{d}t^2$	Newton's second law	$M = \mathrm{d}^2\theta/\mathrm{d}t^2$

3.4 Simple harmonic motion

A particle moves with *simple harmonic motion* when it has constant angular velocity, ω, and follows a displacement pattern

$$x = x_0 \sin\left(\frac{2\pi Nt}{60}\right)$$

The projected displacement, velocity, and acceleration of a point P on the x–y axes are a sinusoidal function of time, t. See Fig. 3.2.

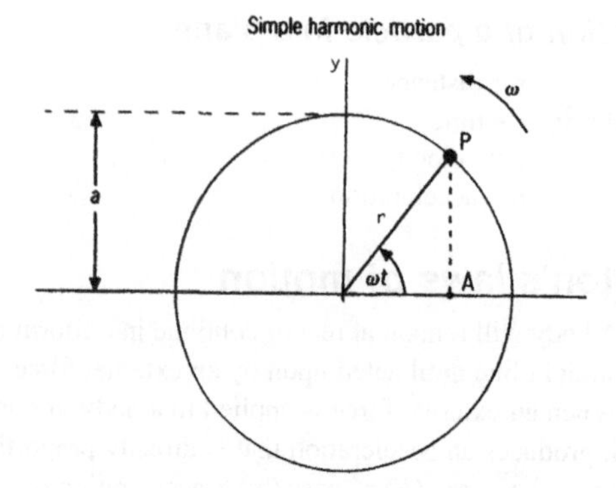

Fig. 3.2 Simple harmonic motion

x_0 = amplitude of the displacement
Angular velocity $\omega = 2\pi N/60$, where N is in r/min
Periodic time $T = 2\pi/\omega$
Velocity, v, of point A on the x axis is

$$v = \mathrm{d}s/\mathrm{d}t = \omega r \sin \omega t$$

Acceleration $a = \mathrm{d}^2s/\mathrm{d}t^2 = \mathrm{d}v/\mathrm{d}t = -\omega^2 r \cos \omega t$

3.5 Understanding acceleration

The dangerous thing about acceleration in rotating components is that it represents a rate of change of speed or velocity. When this rate of change is high, it puts high stresses on the components, causing them to deform and break. In practice, the components of engineering machines experience acceleration many times the force of gravity, so they have to be designed to resist the forces that result. These forces can be caused as a result of either linear or angular accelerations, and there is a comparison between the two as shown below:

Linear acceleration

$$a = \frac{v - u}{t} \text{ m/s}^2$$

Angular acceleration

$$\alpha = \frac{\omega_2 - \omega_1}{t} \text{ rad/s}^2$$

When analysing (or designing) any machine or mechanism, think about linear accelerations first – they are always important.

3.6 Dynamic forces and loadings

The design of rotating equipment is heavily influenced by the need to resist dynamic loads in use.

Dynamic forces can be classified into three main groups:

- suddenly applied loads and simple impact forces;
- forces due to rotating masses;
- forces due to reciprocating masses.

In order to be able to chose design parameters, three factors have to be considered:

- the energy to be absorbed;
- the elastic modulus, *E,* of the material of the impacted member;
- the elastic limit, R_e, or the appropriate fatigue endurance limit, of the material.

A basic 'rule of thumb' equation for impact situations is

$$\sigma^2 = \frac{2EX}{V}$$

where

σ = maximum generated stress
E = Young's modulus of elasticity
X = Energy to be absorbed
V = Effective volume of the impacted member.

This equation uses the basic assumption that the impacted member is infinitely rigidly supported and so absorbs all the energy, hence giving the most severe stress conditions. In practical rotating equipment design factors of approximately three to eight on static stress may be necessary to allow for dynamic loadings.

For a situation where components are subjected to fatigue conditions, the maximum permissible working stress must be adjusted according to the desired life of the structure related to the frequency of the dynamic load cycle. A long-life component (i.e. long life relative to number of cycles, say 10^7) requires an additional safety factor. As a guide, the factor should be equal to at least 2.2 for stresses that fluctuate between zero and a maximum in one direction, and at least 3.2 for stresses operating between equal positive and negative stress maxima (e.g. tension and compression in a shaft rotating under a bending moment).

If a rapid loading or impact cycle is repeated at relatively high uniform frequency, then resonant or harmonic vibration may be set up in a structure, causing severe overloading.

3.7 Forces due to rotating masses

Forces due to rotating masses are another significant factor in rotating equipment design. The two main stresses generated are:

- stresses caused by centrifugal force;
- stresses resulting from inherently unbalanced rotating masses.

A basic formula is

$$\text{Centrifugal force } F = \frac{Wv^2}{gk} = \frac{4Wk\pi^2 n^2}{3600\, g}$$

where

W = weight of revolving body
v = velocity at radius k
g = acceleration due to gravity
n = r/min

The radius of gyration k is defined as the distance from the axis of swing to the centre at which the whole rotating or oscillating mass may be regarded as being concentrated, without involving any change in the moment of inertia. (In this case, this is the true moment of inertia, and not the second moment of area.)

If I is the moment of inertia, then

$$I = \frac{Wk^2}{g}$$

or

$$k = \sqrt{\left(\frac{Ig}{W}\right)}$$

It is unusual for the centre of gyration (i.e. the point at which a mass may be regarded as being concentrated) to coincide with the centre of gravity of the mass, but they do coincide *approximately* if the radial depth of the mass is small compared to the radius of gyration. In such a case, the radius of swing of the centre of gravity may be used for calculation purposes instead of the radius of gyration. A similar reasoning may be applied to calculations for the rim of a wheel if the rim thickness is relatively small and the mass of the rim is regarded as acting through the centroid of its area of cross-section.

3.8 Forces due to reciprocating masses

For simple analysis of rotating masses, it is usually assumed that the reciprocation follows basic simple harmonic motion, see Fig. 3.2.

CHAPTER 4

Rotating Machine Fundamentals: Vibration, Balancing, and Noise

4.1 Vibration: general model

Vibration is a subset of the subject of dynamics. It has particular relevance to both structures and machinery in the way that they respond to applied disturbances. The most common model of vibration is a concentrated spring-mounted mass that is subject to a disturbing force and retarding force, see Fig. 4.1.

The motion is represented graphically as shown by the projection of a rotating vector x. Relevant quantities are

frequency (Hz) = $\sqrt{(k/m)}/2\pi$
k = spring stiffness
m = mass

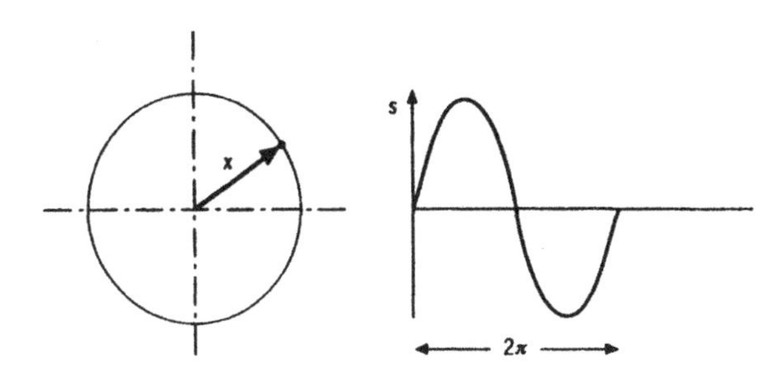

Fig. 4.1 Vibration: the general model

The ideal case represents simple harmonic motion with the waveform being sinusoidal. Hence the motion follows the general pattern:

- vibration displacement (amplitude) = s
- vibration velocity = v = ds/dt
- vibration acceleration = a = dv/dt

4.2 Vibration formulae

The four most common vibration cases are as shown below (see Fig. 4.2).

Free vibration: linear (Fig. 4.2 (a))

$$m\ddot{x} + kx = 0$$

$$x = A \sin(\omega_n t - \phi)$$

$$\omega_n = \sqrt{\left(\frac{k}{m}\right)} = \sqrt{\left(\frac{g}{\Delta}\right)}$$

Free vibration: torsional (Fig. 4.2 (b))

$$J\ddot{\theta} + k_t\theta = 0$$

$$\theta = A \sin(\omega_n t - \phi)$$

$$\omega_n = \sqrt{\left(\frac{k_t}{J}\right)}$$

Free damped vibration (Fig. 4.2 (c))

$$m\ddot{x} + c\dot{x} + kx = 0$$

$$x = \mathrm{A}e^{-\xi\omega_n t} \sin(\omega_d t + \psi)$$

$$\omega_d = \omega_n\sqrt{(1 - \zeta^2)}$$

$$c_c = 2m\omega_n$$

$$\delta = \ln\frac{x_0}{x_1} = \ln\frac{x_1}{x_2} = \frac{2\pi\zeta}{\sqrt{(1-\zeta^2)}}$$

$$m\ddot{x} + c\dot{x} + kx = F \sin \omega t$$

$$x = \mathrm{A}e^{-\xi\omega_n t} \sin(\omega_d t + \psi)$$

$$+ X\sin(\omega t - \phi)$$

Forced vibration with damping (Fig. 4.2 (d))

$$X = \frac{F/k}{\sqrt{[(1-r^2)^2 + (2\zeta r)^2]}}$$

$$D = \frac{1}{\sqrt{[(1-r^2)^2 + (2\zeta r)^3]}} = \frac{X}{X_0}$$

$$\tan \phi = \frac{2\zeta r}{1-r^2}$$

X is maximum when $r = \sqrt{(1 - 2\zeta^2)}$

$$\frac{X_{max}}{X_0} = \frac{1}{2\zeta \sqrt{(1-\zeta^2)}}$$

$$\tan \phi = \frac{1}{\zeta}\sqrt{(1-2\zeta^2)}$$

at resonance, $r = 1$

$$X_{re} = \frac{F}{c\omega_n} = \frac{X_0}{2\zeta}$$

m	mass
k	spring constant
Δ	static deflection
x	displacement
A	constant
ω_n	natural frequency
ϕ, ψ	phase angle
k_t	torsional stiffness of shaft
J	mass moment of inertia of flywheel
θ	angular displacement
ζ	$= c/c_c$ damping factor
c	damping coefficient
c_c	critical damping coefficient
δ	logarithmic decrement
ω_d	natural frequency of damped vibration
F	maximum periodic force
X_0	equivalent static deflection = F/k

X_{max} peak amplitude
X_{re} amplitude at resonance
r $= \omega/\omega_n$, frequency ratio
D dynamic magnifier

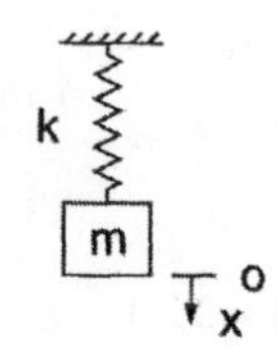

(a) Free vibration (linear)

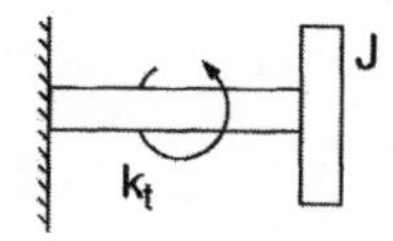

(b) Free vibration (torsional)

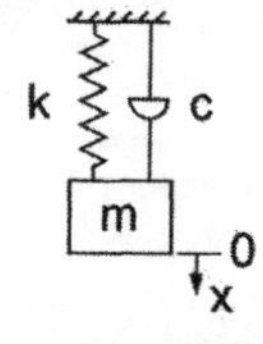

(c) Free damped vibration

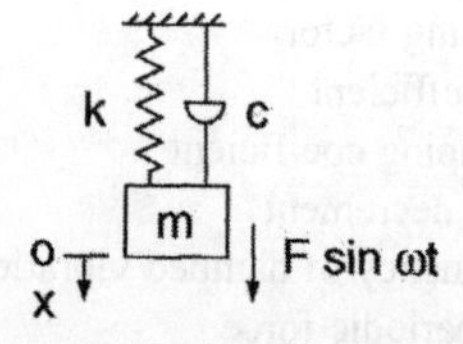

(d) Forced vibration with damping

Fig. 4.2 Vibration modes

4.3 Machine vibration

There are two main types of vibration relevant to rotating machines:

- *bearing housing vibration*. This is assumed to be sinusoidal. It normally uses the velocity (V_{rms}) parameter.
- *shaft vibration*. This is generally not sinusoidal. It normally uses displacement (s) as the measured parameter.

Bearing housing vibration

Relevant points are:

- only vibration at the 'surface' is measured;
- torsional vibration is excluded;
- V_{rms} is normally measured across the frequency range and then distilled down to a single value, i.e. $V_{rms} = \sqrt{[\frac{1}{2}\Sigma(\text{amplitudes} \times \text{angular frequences})]}$.

Acceptance levels

Technical standards and manufacturers' practices differ in their acceptance levels. General 'rule of thumb' acceptance levels are shown in Tables 4.1 and 4.2, and Fig. 4.3.

Table 4.1 Balance quality grades (ISO 1940)

Balance quality grade G	*eω* (mm/s)*	*Rotor types – general examples*
G 4000	4000	Crankshaft drives of rigidly mounted, slow marine diesel engines with uneven number of cylinders
G 1600	1600	Crankshaft drives of rigidly mounted, large, two-cycle engines
G 630	630	Crankshaft drives of rigidly mounted, large, four-cycle engines
		Crankshaft drives of elastically mounted marine diesel engines
G 250	250	Crankshaft drives of rigidly mounted, fast, four-cylinder diesel engines
G 100	100	Crankshaft drives of fast diesel engines with six or more cylinders
		Complete engines (gasoline or diesel) for cars, trucks, and locomotives
G 40	40	Car wheels, wheel rims, wheel sets, drive shafts
		Crankshaft drives of elastically mounted, fast, four-cycle engines (gasoline or diesel) with six or more cylinders

Table 4.1 Cont.

		Crankshaft drives for engines of cars, trucks, and locomotives
G 16	16	Drive shafts (propeller shafts, cardan shafts) with special requirements
		Parts of crushing machinery
		Parts of agricultural machinery
		Individual components of engines (petrol or diesel) for cars, trucks, and locomotives
		Crankshaft drives of engines with six or more cylinders under special requirements
G 6.3	6.3	Parts of process plant machines
		Marine main turbine gears
		Centrifuge drums
		Fans
		Assembled aircraft gas turbine rotors
		Flywheels
		Pump impellers
		Machine-tool and general machinery parts
		Normal electrical armatures
		Individual components of engines under special requirements
G 2.5	2.5	Gas and steam turbines, including marine main turbines.
		Rigid turbogenerator rotors
		Rotors
		Turbocompressors
		Machine-tool drives
		Medium and large electrical armatures with special requirements
		Small electrical armatures
		Turbine-driven pumps
G 1	1	Tape recorder and phonograph (gramophone) drives
		Grinding-machine drives
		Small electrical armatures with special requirements
G 0.4	0.4	Spindles, disks, and armatures of precision grinders
		Gyroscopes

*$\omega = 2\pi \times N/60 \propto n/10$, if n is measured in r/min and ω in rad/s. e is the eccentricity of the centre of gravity.

Table 4.2 General 'rules of thumb' acceptance levels

Machine	V_{rms} *(mm/s)*
Precision components and machines – gas turbines, etc.	1.12
Helical and epicyclic gearboxes	1.8
Spur-gearboxes, turbines	2.8
General service pumps	4.5
Long-shaft pumps	4.5–7.1
Diesel engines	7.1
Reciprocating large machines	7.1–11.2

Typical balance grades; from ISO 1940–1

Balance grade	*Type of rotor (general examples)*
G 1	Grinding machines, tape-recording equipment
G 2.5	Turbines, compressors, electric armatures
G 6.3	Pump impellers, fans, gears, machine tools
G 16	Cardan shafts, agriculture machinery
G 40	Car wheels, engine crankshafts
G 100	Complete engines for cars and trucks

'Acceptance criteria': from ISO 10816-1

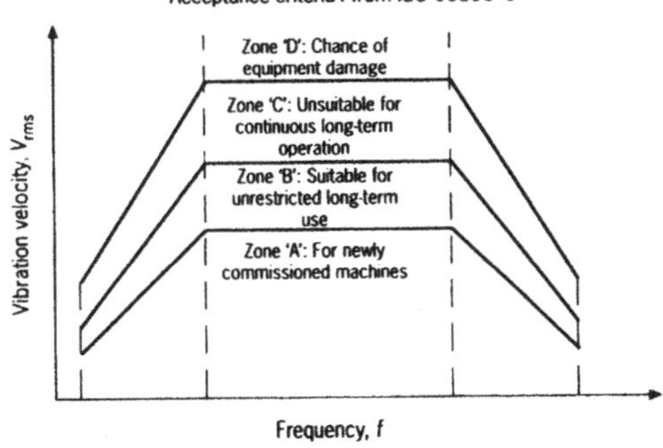

Typical 'boundary limits': from ISO 10816-1

V_{rms}	Class I	Class II	Class III	Class IV
0.71	A	A	A	A
1.12	B	A	A	A
1.8	B	B	A	A
2.8	C	B	B	A
4.5	C	C	B	B
7.1	D	C	C	B
11.2	D	D	C	C
18	D	D	D	C

Class suitability

- Class I — Machines < 15 kW
- Class II — Machines < 300 kW
- Class III — Large machines with rigid foundations
- Class IV — Large machines with 'soft' foundations

(Note how wide these classes are)

Fig. 4.3 Vibration balance grades ISO 10816-1

4.4 Dynamic balancing

Almost all rotating machines (pumps, shafts, turbines, gearsets, generators, etc.) are subject to dynamic balancing during manufacture. The objective is to maintain the operating vibration of the machine within manageable limits.

Dynamic balancing normally comprises two measurement/correction planes and involves the calculation of vector quantities. The component is mounted in a balancing rig which rotates it at near its operating speed, and both senses and records out-of-balance forces and phase angle in two planes. Balance weights are then added (or removed) to bring the imbalance forces to an acceptable level (see Fig. 4.4). Figure 4.5 shows how to interpret the corresponding vibration readings.

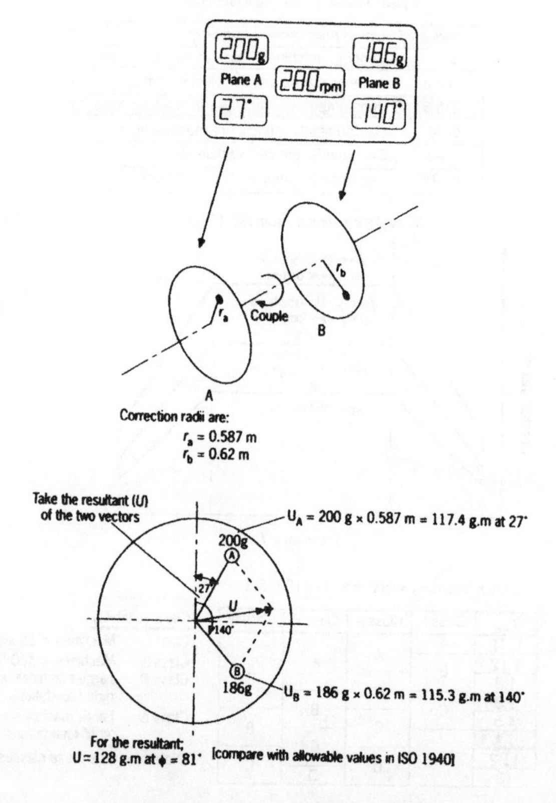

Fig. 4.4 Dynamic balancing

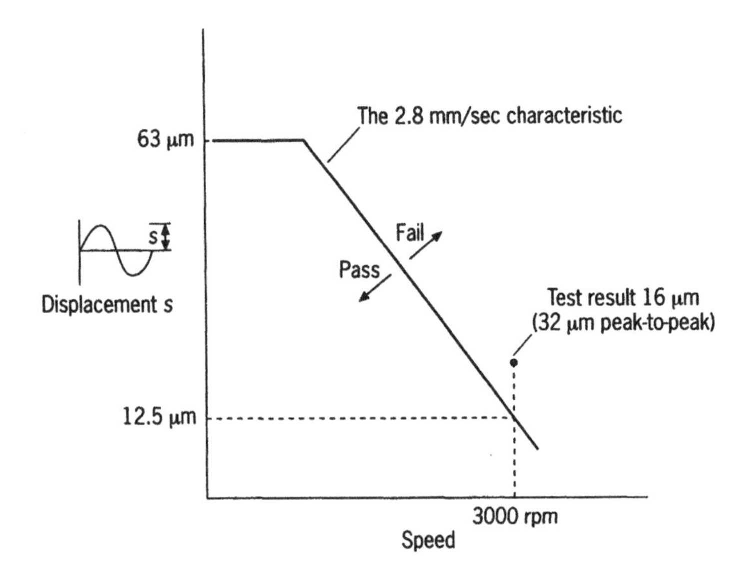

Fig. 4.5 How to interpret vibration readings

Balancing standards

The international standards ISO 1940-1 (1984) *Balance and quality requirements of rigid rotors* and ISO 10816-1 are frequently used. Finer balance grades are used for precision assemblies such as instruments and gyroscopes. The nearest American equivalent is ANSI/ASA standard ANSI S2.42 (1997) *Balancing of flexible rotors*. This also classifies rotors into groups in accordance with various balance 'quality' grades.

4.5 Machinery noise

Principles

Noise is most easily thought of as airborne pressure pulses set up by a vibrating surface source. It is measured by an instrument that detects these pressure changes in the air and then relates this measured sound pressure to an accepted *zero* level. Because a machine produces a mixture of frequencies (termed 'broad-band' noise), there is no single noise measurement that will fully describe a noise emission. In practice, two methods used are:

- *The 'overall noise' level.* This is often used as a colloquial term for what is properly described as the *A-weighted sound pressure level.* It incorporates multiple frequencies, and weights them according to a formula that results in the best approximation of the loudness of the noise. This is displayed as a single instrument reading expressed as decibels dB(A).
- *Frequency band sound pressure level.* This involves measuring the sound pressure level in a number of frequency bands. These are arranged in either octave or one-third octave bands in terms of their mid-band frequency. The range of frequencies of interest in measuring machinery noise is from about 30 Hz to 10 000 Hz. Note that frequency band sound pressure levels are also expressed in decibels (dB).

The decibel scale itself is a logarithmic scale – a sound pressure level in dB being defined as

$$dB = 10 \log_{10} (p_1/p_0)^2$$

where

p_1 = measured sound pressure
p_0 = a reference *zero* pressure level

Noise tests on rotating machines are carried out by defining a 'reference surface' and then positioning microphones at locations 3 ft (0.91 m) from it (see Fig. 4.6).

Typical levels

Approximate 'rule of thumb' noise levels are given in Table 4.3.

Table 4.3 Typical noise levels

Machine/environment	dB(A)
A whisper	20
Office noise	50
Noisy factory	90
Large diesel engine	97
Turbocompressor/gas turbine	98

A normal 'specification' level is 90–95 dB(A) at 1 m from operating equipment. Noisier equipment needs an acoustic enclosure. Humans can continue to hear increasing sound levels up to about 120 dB. Levels above this cause serious discomfort and long-term damage.

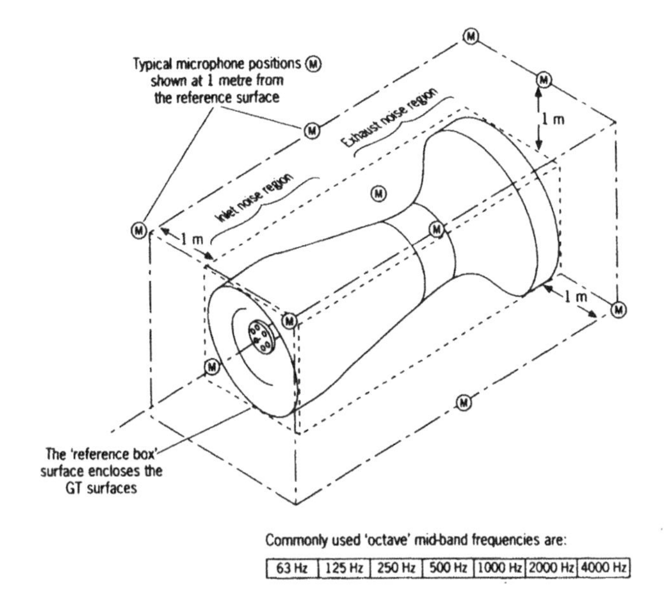

Fig. 4.6 Noise tests on rotating machines

4.6 Useful references

Standards: balancing

1. API publication 684: (1992) First edition, *A tutorial on the API approach to rotor dynamics and balancing.*
2. SAE ARP 5323: (1988) *Balancing machines for gas turbine rotors.*

Standards: vibration

Table 4.4 shows the status of some relevant technical standards dealing with vibration.

Table 4.4 Technical standards – vibration

Standard	*Title*	*Status*
BS 4675-2: 1978, ISO 2954-1975	Mechanical vibration in rotating machinery. Requirements for instruments for measuring vibration severity.	Current
CP 2012-1: 1974	Code of practice for foundations for machinery. Foundations for reciprocating machines.	Current
BS EN 1032: 1996	Mechanical vibration. Testing of mobile machinery in order to determine the whole-body vibration emission value. General.	Current Work in hand
BS EN 12786: 1999	Safety of machinery. Guidance for the drafting of vibration clauses of safety standards.	Current
00/710581 DC	ISO/DIS 14839-1 Mechanical vibration of rotating machinery equipped with active magnetic bearings. Part 1. Vocabulary.	Current Draft for public comment
BS 4675: Part 1: 1976, ISO 2372-1974	Mechanical vibration in rotating machinery. Basis for specifying evaluation standards for rotating machines with operating speeds from 10 to 200 rev/s.	Withdrawn Superseded

See also Table 13.11 showing harmonized standards relevant to the machinery directive.

Standards: noise

1. ANSI/ASA S12.16: (1997) *American National Standard Guidelines for the specification of noise from new machinery.*
2. ANSI/ASA S12.3: (1996) *American National Standard Statistical methods for determining and verifying stated noise emission values of machinery and equipment.*
3. ISO 10494: (1993) *Gas turbine and gas turbine sets – measurement of emitted airborne noise – engineering (survey method).*

CHAPTER 5

Machine Elements

'Machine elements' is the term given to the set of basic mechanical components that are used as building blocks to make up a rotating equipment product or system. There are many hundreds of these; the most common ones are shown, subdivided into their common groupings, in Fig. 5.1. The established reference source for the design of machine elements is:

- Shigley, J.E. and Mischke, C.R. (1996) *Standard handbook of machine design*, Second edition, McGraw Hill, ISBN 0-07-056958-4.

5.1 Screw fasteners

The ISO metric and, in the USA, the unified inch and ISO inch are commonly used in rotating equipment designs. They are covered by different technical standards, depending on their size, material, and application.

Dimensions: ISO metric fasteners (ISO 4759)

Table 5.1 and Fig. 5.2 show typical dimensions (all in mm) for metric fasteners covered by ISO 4759.

Table 5.1 ISO metric fastener dimensions (mm)

Size	*Pitch*	*Width A/F (F)*		*Head height (H)*		*Nut thickness (m)*	
		max	*min*	*max*	*min*	*max*	*min*
M5	0.8	8.00	7.85	3.650	3.350	4.00	3.7
M8	1.25	13.00	12.73	5.650	5.350	6.50	6.14
M10	1.5	17.00	16.73	7.180	6.820	8.00	7.64
M12	1.75	19.00	18.67	8.180	7.820	10.00	9.64
M20	2.5	30.00	29.67	13.215	12.785	16.00	15.57

Locating	*Drives and mechanisms*	*Energy transmission*	*Rotary bearings*	*Dynamic sealing*
Threaded fasteners	**Shafts**	**Gear trains**	**Rolling**	**Rotating shaft seals**
Nuts and bolts	Parallel	Spur	Ball	Face
Set screws	Taper	Helical	Roller (parallel)	Interstitial
Studs	Concentric	Bevel	Roller (tapered)	Axial radial
Grub screws	**Mechanisms**	Worm and wheel	Needle	Bush
Expanding bolts	Crank and sliding	Epicyclic	Self-aligning	Labyrinth
Keys	Ratchet and pawl	**Belt drives**	**Sliding**	Lip ring
Flat	Geneva	Flat	Axial	Split ring
Taper	Scotch-yolk	Vee	Radial	**Reciprocating shaft seals**
Woodruff	Carden joint	Wedge	Bush	Piston rings
Profiled	**Cams**	Synchronous	Hydrodynamic	Packing rings
Pins	Constant velocity	**Chain drives**	Hydrostatic	
Split	Uniform acceleration	Roller	Self-lubricating	
Taper	Simple harmonic	Conveyor	Slideways	
Splines	Motion (s.h.m.)	Leaf		
Retaining rings	**Clutches**	**Pulleys**		
Clamps	Dog	Simple		
Clips	Cone	Differential		
Circlips	Disc	**Springs**		
Spring	Spring	Tension		
Shoulders and grooves	Magnetic	Compression		
	Fluid coupling			
	Brakes			
	Disk			
	Drum			
	Couplings			
	Rigid			
	Flexible			
	Spring			
	Membrane			
	Cordon			
	Claw			

Fig. 5.1 Machine elements

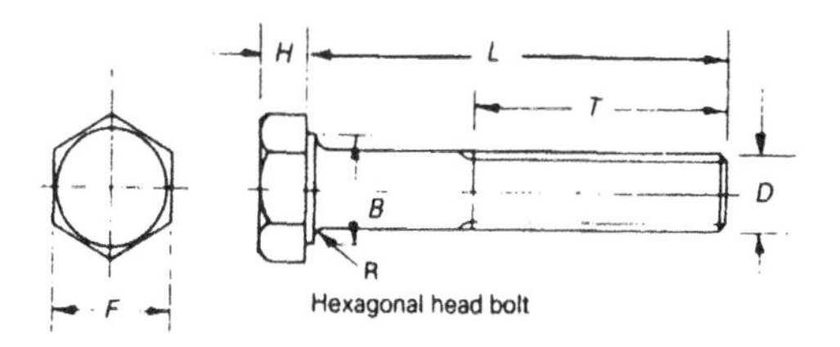

Fig. 5.2 Fastener dimensions

Unified inch screw threads (ASME B1.1)

Fasteners are defined by their combination of diameter–pitch relationship and tolerance class. Table 5.2 shows the system of unified inch thread designation (see also Table 5.3 for UNC/UNRC thread dimensions).

Table 5.2 Unified inch thread relationships

Diameter–pitch relationship	*Tolerance class*
UNC and UNRC: Coarse	A represents external threads
UNF and UNRF: Fine	B represents internal threads
UN and UNR: Constant pitch	Class 1: Loose tolerances for easy assembly
UNEF and UNREF: Extra fine	Class 2: Normal tolerances for production items
	Class 3: Close tolerances for accurate location application

Table 5.3 Typical* UNC/UNRC thread dimensions

Nominal size (in)	*Basic major diameter D (in)*	*Threads per inch (in)*	*Basic minor diameter K (in)*
1/8	0.125	40	0.0979
1/4	0.25	20	0.1959
1/2	0.50	13	0.4167
1	1.00	8	0.8647
1½	1.50	6	1.3196
2	2.00	4½	1.7594

*Data from ANSI B1.1: 1982. Equivalent to ISO 5864: 1993.

ISO metric screw threads (ISO 261)

The ISO thread profile is similar to the unified screw thread. They are defined by a set of numbers and letters as shown in Fig. 5.3.

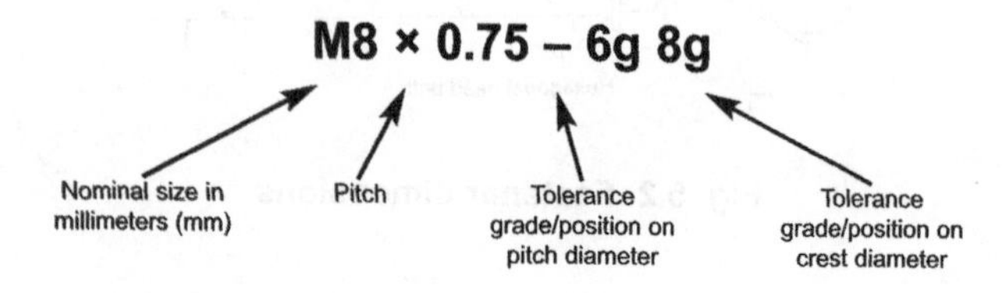

Fig. 5.3 Typical ISO metric thread designation

5.2 Bearings

Types

Bearings are basically subdivided into three types: sliding bearings (plane motion), sliding bearings (rotary motion), and rolling element bearings (see Fig. 5.4). There are three lubrication regimes for sliding bearings:

- *boundary lubrication*: there is actual physical contact between the surfaces;
- *mixed-film lubrication*: the surfaces are partially separated for intermittent periods;
- *full-film 'hydrodynamic' lubrication*: the two surfaces 'ride' on a wedge of lubricant.

Ball and roller bearings

Some of the most common designs of ball and roller bearings are shown in Fig. 5.5. The amount of misalignment that can be tolerated is a critical factor in design selection. Roller bearings have higher basic load ratings than equivalent-sized ball types.

Bearing lifetime

Bearing lifetime ratings are used in purchasers' specifications and manufacturers' catalogues and datasheets. The rating life (L_{10}) is that corresponding to a 10 per cent probability of failure and is given by:

L_{10} radial ball bearings $= (Cr/Pr)^3 \times 10^6$ revolutions
L_{10} radial roller bearings $= (Cr/Pr)^{10/3} \times 10^6$ revolutions
L_{10} thrust ball bearings $= (Ca/Pa)^3 \times 10^6$ revolutions
L_{10} thrust roller bearings $= (Ca/Pa)^{10/3} \times 10^6$ revolutions

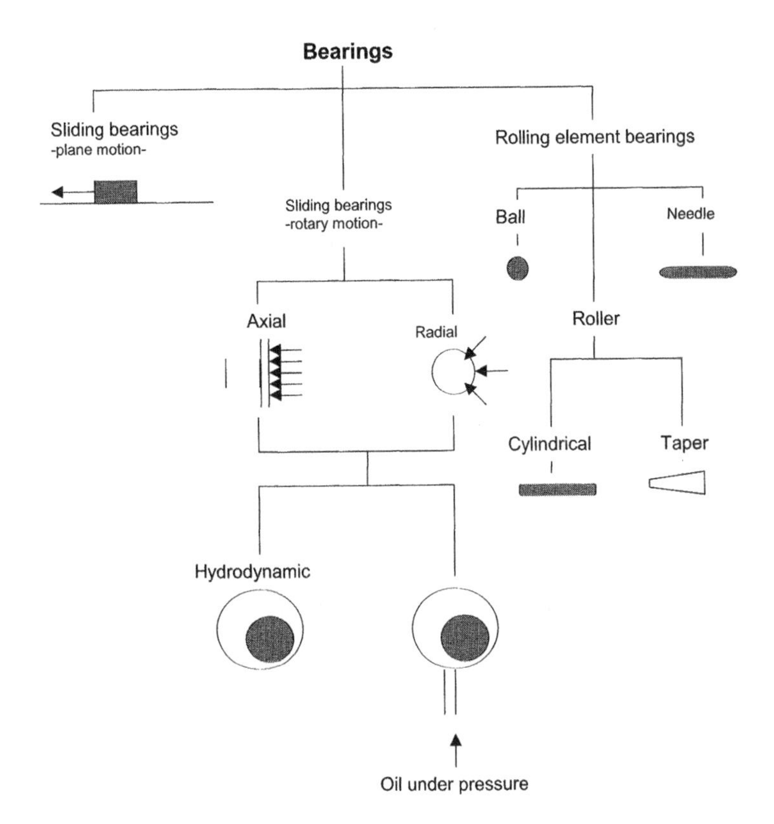

Fig. 5.4 Bearing types

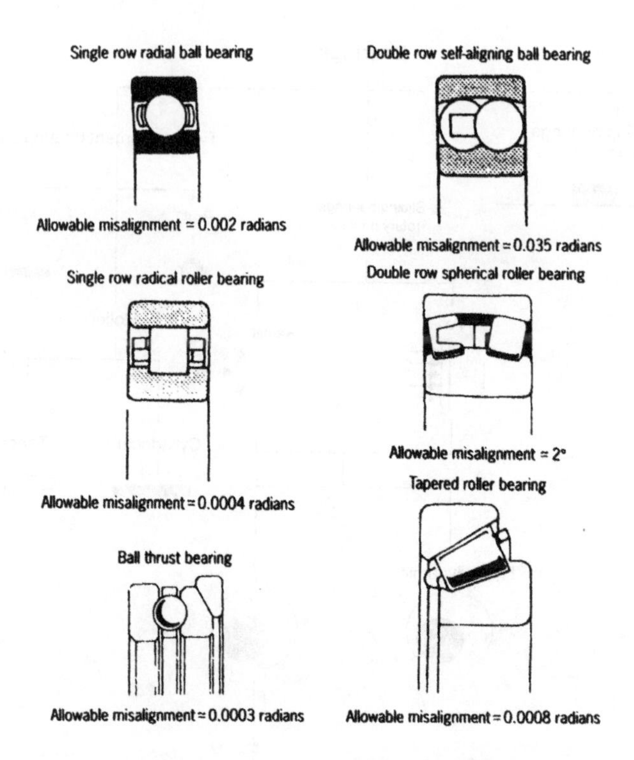

Fig. 5.5 Ball and roller bearings

Cr and Ca are the static radial and axial load ratings that the bearing can theoretically endure for 10^6 revolutions. Pr and Pa are corresponding dynamic equivalent radial and axial loads.

So, as a general case:

roller bearings: L_{10} lifetime $= [16700\ (C/P)^{10/3}]/n$
ball bearings: L_{10} lifetime $= [16700(C/P)^3]/n$

where

C = Cr or Ca } as appropriate
P = Pr or Pa }
n = speed in r/min

Coefficients of friction

The coefficient of friction between bearing surfaces is an important design criterion for machine elements that have rotating, meshing, or mating parts. The coefficient value (f) varies, depending on whether the surfaces are static or already sliding, and whether they are dry or lubricated. Table 5.4 shows some typical values.

Table 5.4 Typical friction (f) coefficients

	Static (f_o)		*Sliding (f)*	
Material	***Dry***	***Lubricated***	***Dry***	***Lubricated***
Steel/steel	0.75	0.15	0.57	0.10
Steel/cast iron	0.72	0.20	0.25	0.14
Steel/phosphor bronze	–	–	0.34	0.18
Steel/bearing 'white metal'	0.45	0.18	0.35	0.15
Steel/tungsten carbide	0.5	0.09	–	–
Steel/aluminium	0.6	–	0.49	–
Steel/Teflon	0.04	–	–	0.04
Steel/plastic	–	–	0.35	0.06
Steel/brass	0.5	–	0.44	–
Steel/copper	0.53	–	0.36	0.2
Steel/fluted rubber	–	–	–	0.05
Cast iron/cast iron	1.10	–	0.15	0.08
Cast iron/brass	–	–	0.30	–
Cast iron/copper	1.05	–	0.30	–
Cast iron/hardwood	–	–	0.5	0.08
Cast iron/zinc	0.85	–	0.2	–
Hardwood/hardwood	0.6	–	0.5	0.17
Tungsten carbide/tungsten carbide	0.2	0.12	–	–
Tungsten carbide/steel	0.5	0.09	–	–
Tungsten carbide/copper	0.35	–	–	–

Note: The static friction coefficient between similar materials is high, and can result in surface damage or seizure.

5.3 Mechanical power transmission – broad guidelines

Because of the large variety of types of rotating equipment that exist, the basic characteristics of such equipment can vary greatly. One of the main ways of classifying such equipment is by reference to its speed/torque/power characteristics. Large heavy-duty machines such as kilns, crushers, etc. have low-speed, high-torque characteristics whereas small precision equipment lies at the opposite end of the scale, producing low torque at high rotational speeds. Figure 5.6 shows some guidelines on the speed/power/torque characteristics of a broad range of rotating equipment types.

Machines that require a high-torque, low-speed output have to be matched to their higher speed prime mover (diesel engine, electric motor, etc.) by a speed reduction mechanism such as gears, chain/belt, or hydraulic drive. Table 5.5 shows some guidelines on the characteristics of various types.

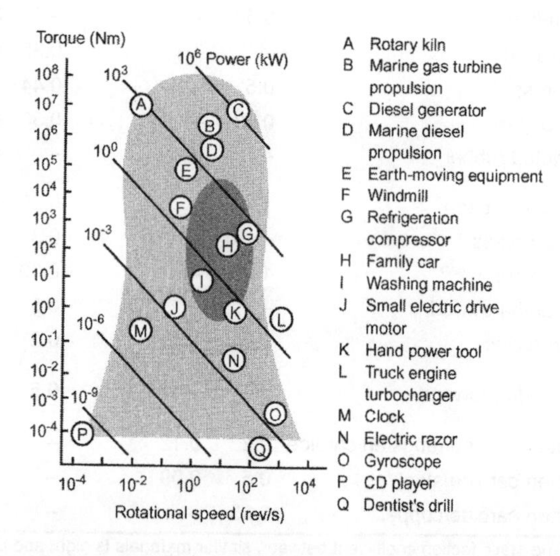

Fig. 5.6 Torque:speed relationships – broad guidelines

Table 5.5 Speed reduction/torque increase mechanisms – broad guidelines

Characteristic	*Gear drive*	*Chain drive*	*Belt drive*	*Hydraulic drive*
Maximum speed	60 m/s	14–17 m/s	25–60 m/s	–
Maximum power capacity	16–18 MW	600 kW	1200 kW	1200–1600 kW
Maximum torque	10^8 Nm	10^6 Nm	10^4 Nm	10^8 Nm

As a general 'rule of thumb', gear drives are more efficient than belt drives and suffer less from vibration problems, but they are much less tolerant to manufacturing inaccuracies and misalignment. Both gear and belt drives provide a speed reduction ratio equivalent to the ratio of the radii of the drive pair.

5.4 Shaft couplings

Shaft couplings are used to transfer drive between two (normally co-axial) shafts. They allow rigid, or slightly flexible, coupling depending on the application. Figure 5.7 shows a typical 'application chart' for several common types.

Bolted couplings

The flanges are rigidly connected by bolts, allowing virtually no misalignment. Positive location may be achieved by using a spigot on the flange face (see Fig. 5.8).

Bushed pin couplings

Similar to the normal bolted coupling, but incorporating rubber bushes in one set of flange holes. This allows a limited amount of angular misalignment (see Fig. 5.9).

Gear couplings

Gear couplings consist of involute-toothed hubs which mesh with an intermediate sleeve or shaft (see Fig. 5.10). They allow significant amounts of angular misalignment and axial movement. Figure 5.11 shows a typical performance envelope, demonstrating the operational limitations of rotational speed, power transmitted, gear tooth contact stress, and centrifugal stress.

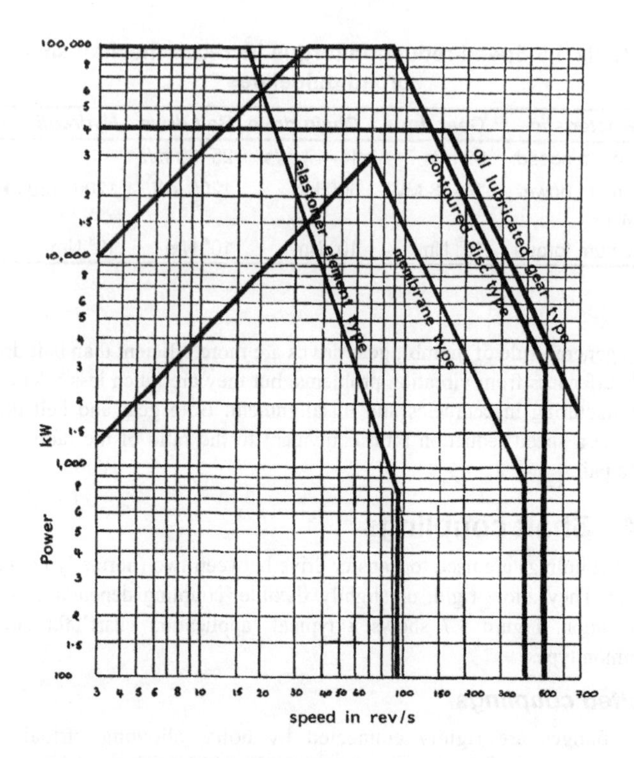

Fig. 5.7 Application chart of coupling types by factored power and speed

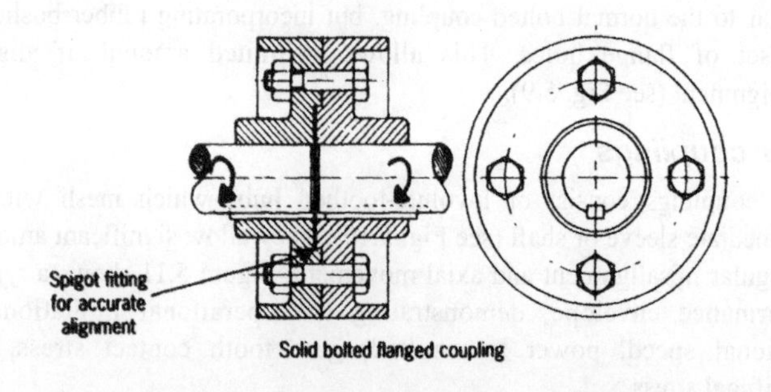

Fig. 5.8 Solid bolted flange coupling

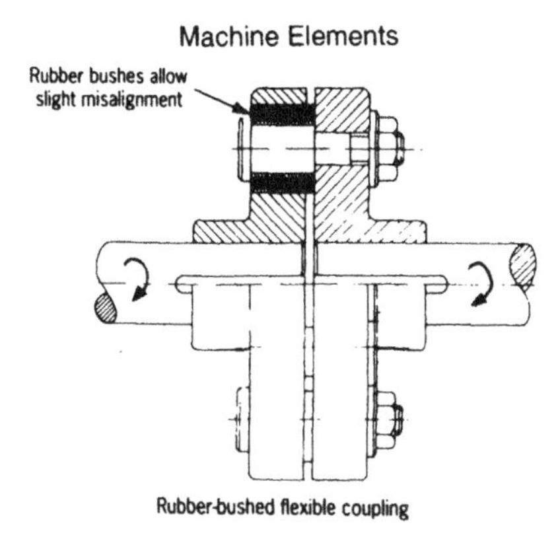

Fig. 5.9 Rubber-bushed coupling

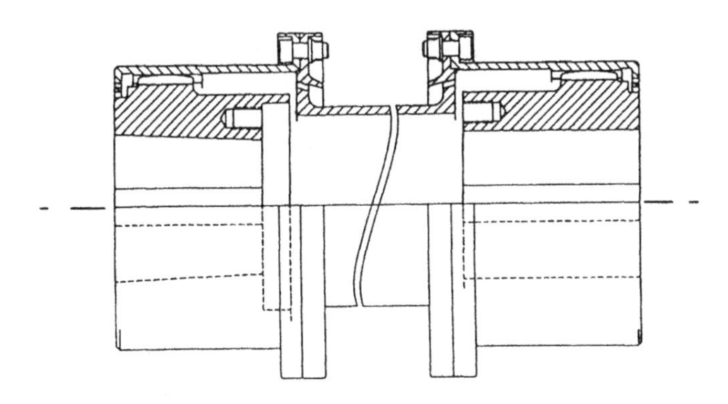

Fig. 5.10 A typical gear coupling

Simple disc-type flexible couplings

A rubber disc is bonded between thin steel discs held between the flanges (see Fig. 5.12).

Membrane-type flexible couplings

These are used specifically for high-speed drives such as gas turbine gearboxes, turbocompressors, and pumps. Two stacks of flexible steel

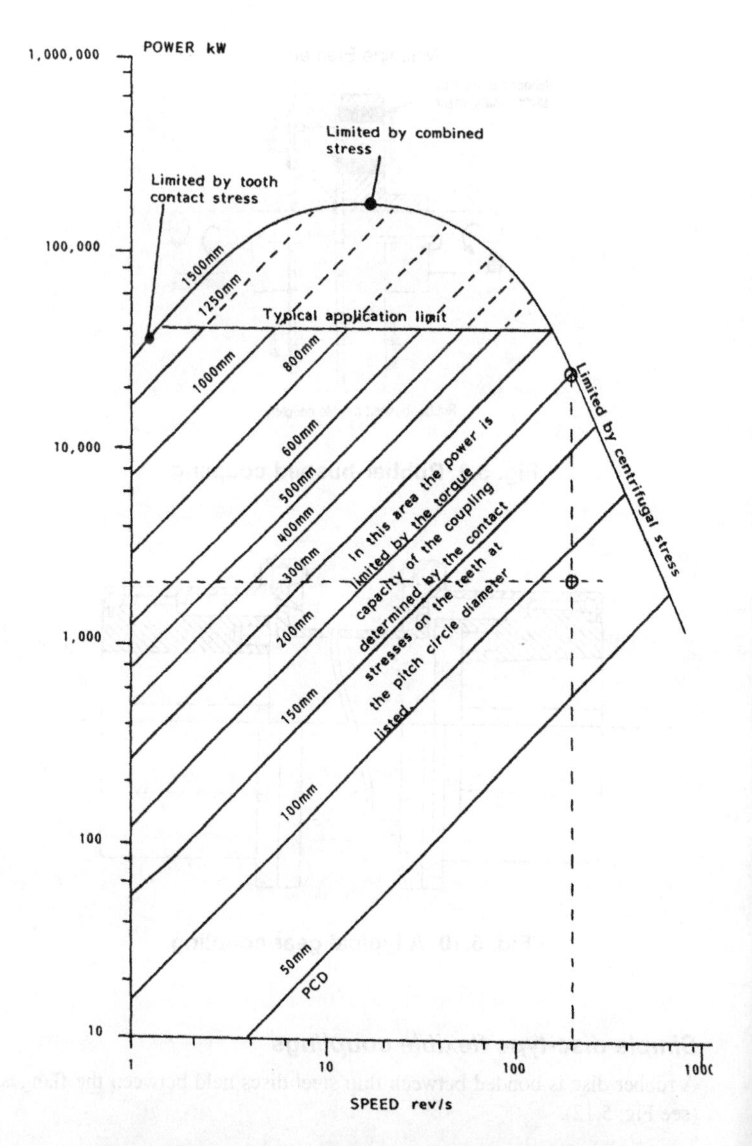

Fig. 5.11 Gear coupling performance envelope

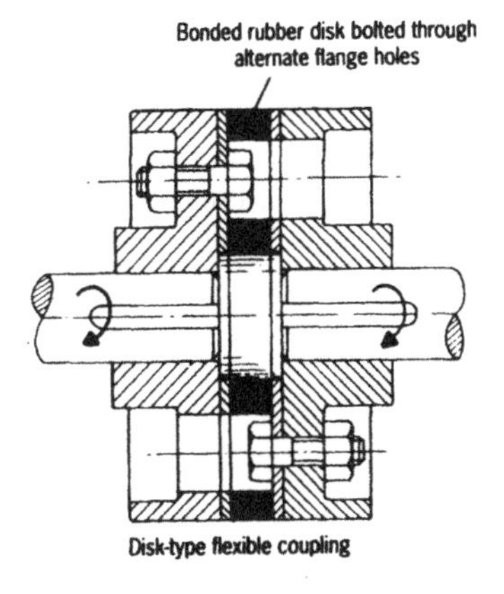

Fig. 5.12 Disc-type flexible coupling

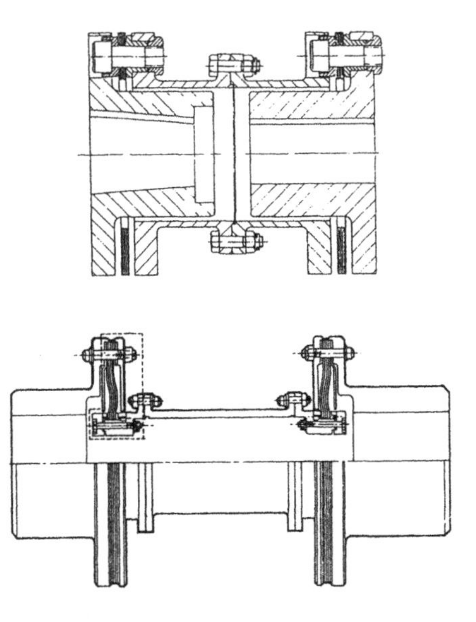

Fig. 5.13 Typical membrane-type flexible couplings

diaphragms fit between the coupling and its mating input/output flanges. These diaphragms are flexible in bending, but strong in tension and shear. These couplings are installed with a static prestretch – the resultant axial force varies with rotating speed and operating temperature. Figure 5.13 shows two common designs. The performance of these couplings is also limited by centrifugal stress considerations. Figure 5.14 shows typical performance envelopes.

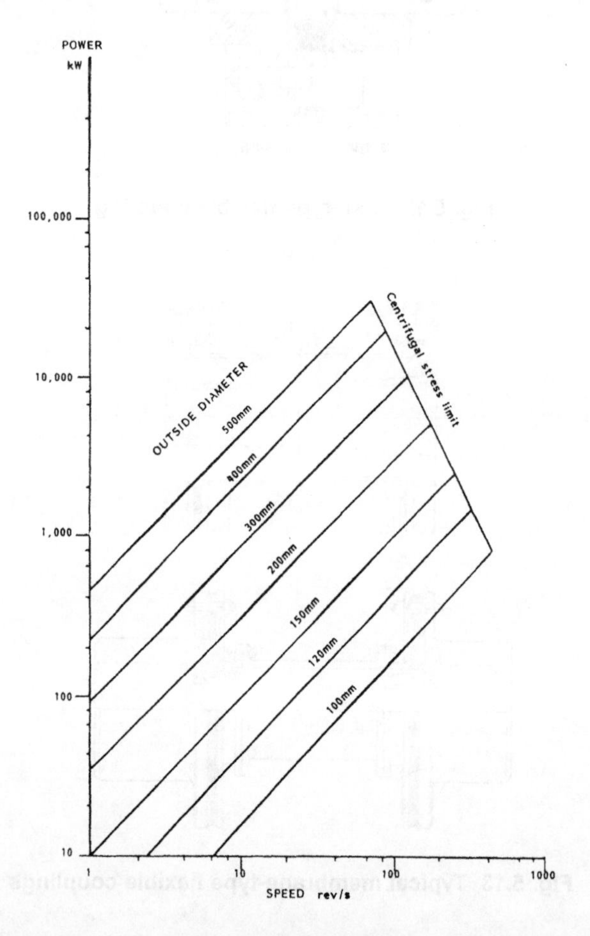

Fig. 5.14 Performance envelopes for membrane-type flexible couplings

Complex-disc couplings

These are a more complicated version of the simple disc-type flexible coupling, used in higher speed applications. Two sets of flexible discs are fitted at either end of a central spacer tube (see Fig. 5.15). Figure 5.16 shows typical performance envelopes which exhibit centrifugal stress limitations.

Balance of couplings

High-speed (and many low-speed) couplings need to be balanced to minimize vibration effects. Table 5.6 shows some typical rules of thumb.

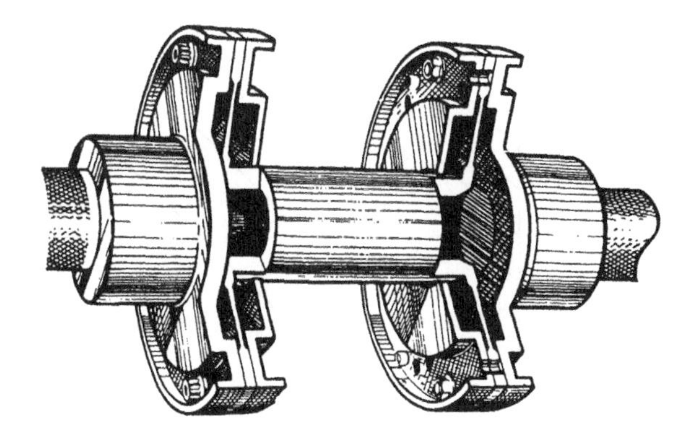

Fig. 5.15 Contoured disc coupling

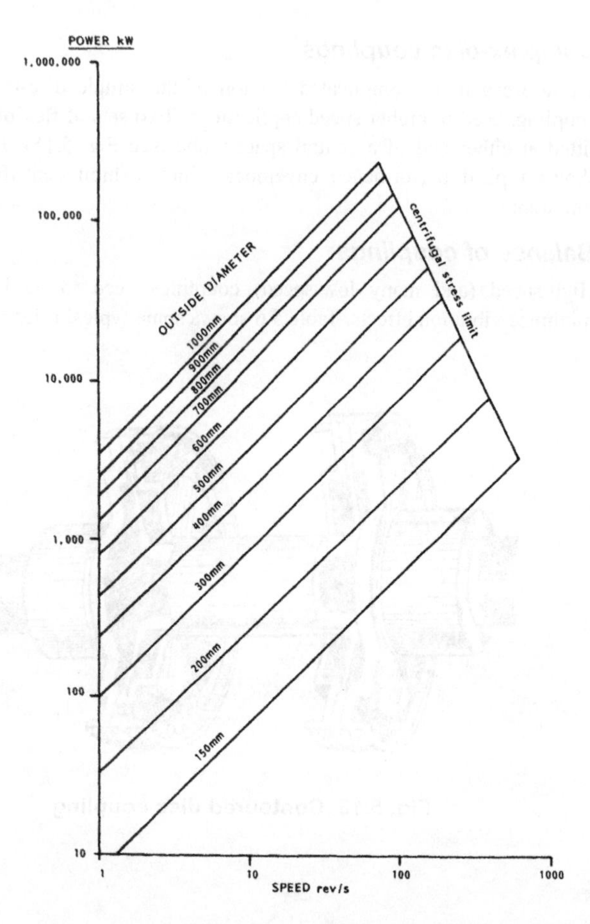

Fig. 5.16 Performance envelopes for contoured disc coupling

Table 5.6 Balance of couplings – some rules of thumb

- API 671 contains guidance on balance for 'rigid' couplings.
- A rigid coupling is one in which shaft bending does not significantly affect the balance – it runs well below the first critical speed (i.e. $N/N_c < 0.2$ approximately).
- For non-API couplings, ISO 1940 is normally used (see Table 4.1). Balance grade G16 is in common use for low-speed 'rigid' couplings.
- For flexible couplings that are defined as sub-critical, i.e. operating speed/first critical speed $N/N_c >$ about 0.2, the ISO 1940 balance grade should be increased.

5.5 Gears

Gear trains are used to transmit motion between shafts. Gear ratios and speeds are calculated using the principle of relative velocities. The most commonly used arrangements are simple or compound trains of spur or helical gears, epicyclic, and worm and wheel.

Simple trains

Simple trains have all their teeth on their 'outside' diameter (see Fig. 5.17).

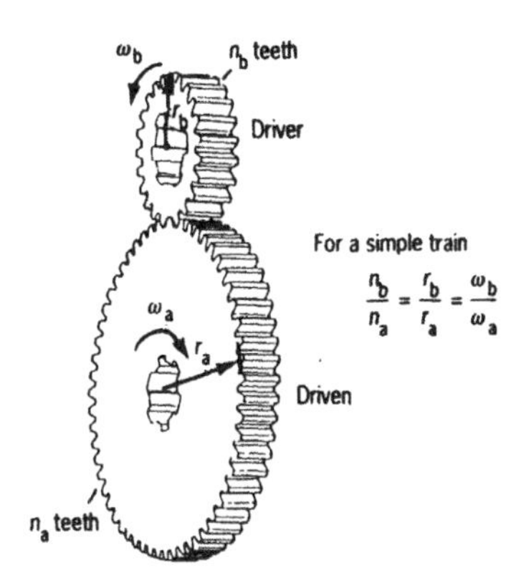

Fig. 5.17 Simple gear train

Compound trains

Speeds are calculated as shown in Fig. 5.18.

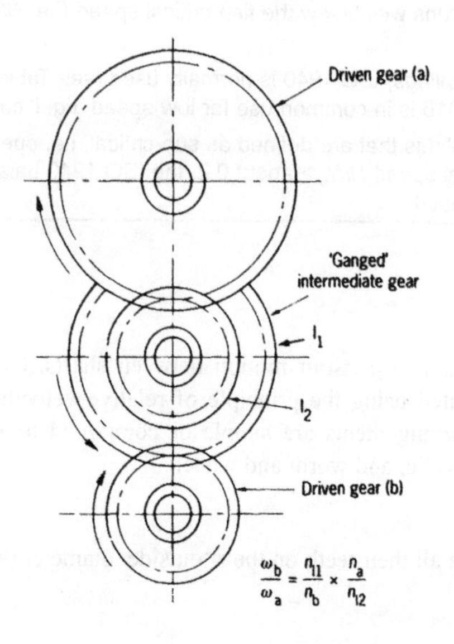

Fig. 5.18 Compound gear train

Worm and wheel

A worm and wheel arrangement is used to transfer drive through 90 degrees, usually incorporating a high gear ratio and output torque. The wheel is a helical gear, see Fig. 5.19.

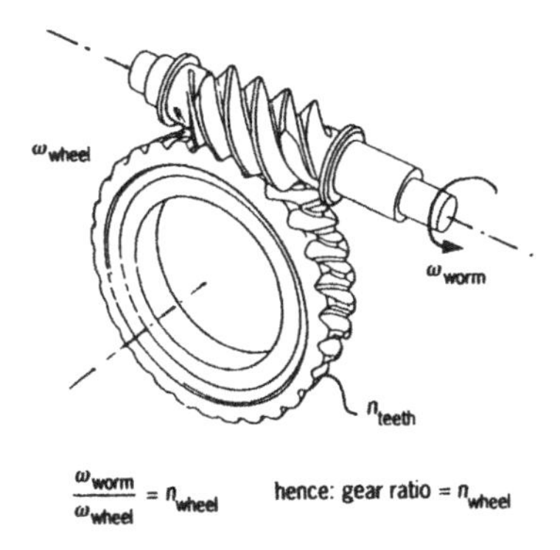

Fig. 5.19 Worm and wheel

Spur gears

Spur gears are the simplest form of gearing arrangement used to transmit power between shafts rotating at (usually) different speeds (see Fig. 5.20). In most applications, spur gear sets are used for speed reduction, i.e. the power is transmitted from a high-speed, low-torque input shaft to a low-speed, high-torque output shaft. Compound trains may also be used.

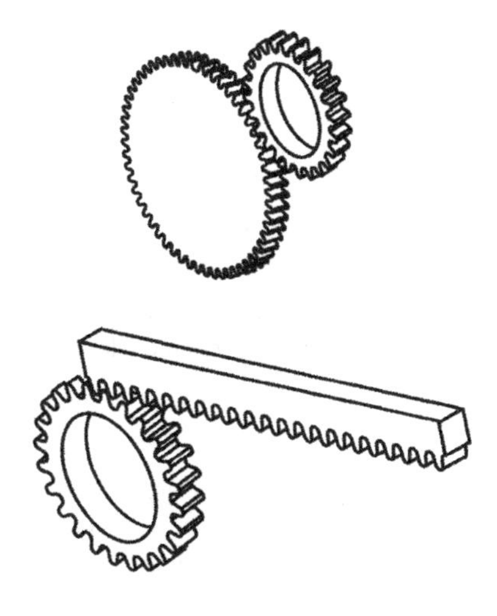

Fig. 5.20 Spur gears

Tooth geometry and kinetics

Spur gear teeth extend from the root or 'dedendum' circle to the tip or 'addendum' circle (see Fig. 5.21). The 'face' or 'flank' is the portion of the tooth that provides the 'drive' contact to the mating gear. The root region contains a fillet to reduce fatigue stresses, and a root clearance.

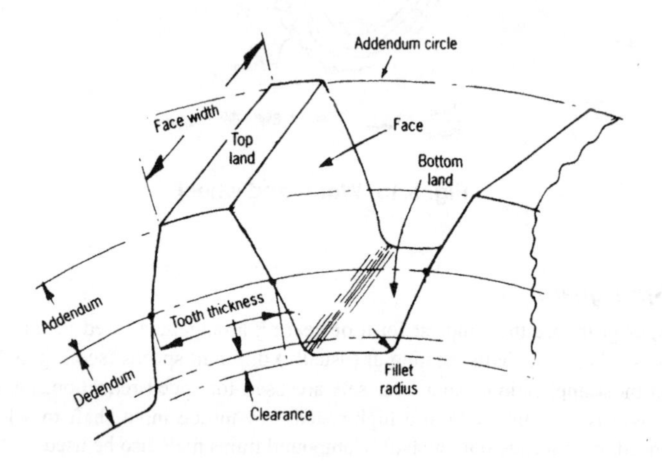

Fig. 5.21 Tooth geometry

For kinetic analysis purposes, spur gears are regarded as equivalent pitch cylinders which roll against each other without any slip. Note that the pitch cylinder diameter is a 'theoretical' dimension.

Other key parameters for spur gear sets are:

- Circular pitch — p = distance between adjacent teeth around the pitch circle

  z = number of teeth on a gear of pitch diameter D
- Module — m = a measure of size = p/π; the module must be the same for both gears in a meshing set
- Pitch point — P = a 'theoretical' point at which the pitch circles of the gears contact each other

- Pitch line velocity — v = the velocity of the pitch point, P
- Tangential force component — P_t = the tangential force component at the pitch point P resulting from the meshing contact between the gears; it is this force component that transfers the power
- Radial force component — P_r = the radial force component (i.e. plays no part in the power transfer)
- Axial force component — P_a = force acting axially along the direction of the gear shaft; it is zero for spur gears

Double helical gears

These are used in most high-speed gearboxes. The double helices produce opposing axial forces that cancel each other out (see Figs 5.22 and 5.23).

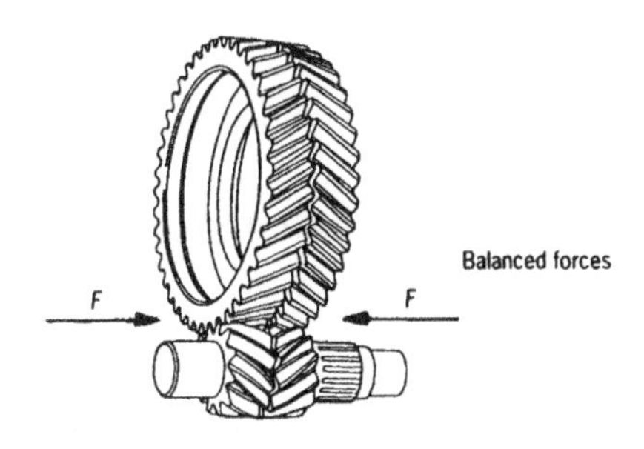

Fig. 5.22 Double helical gears

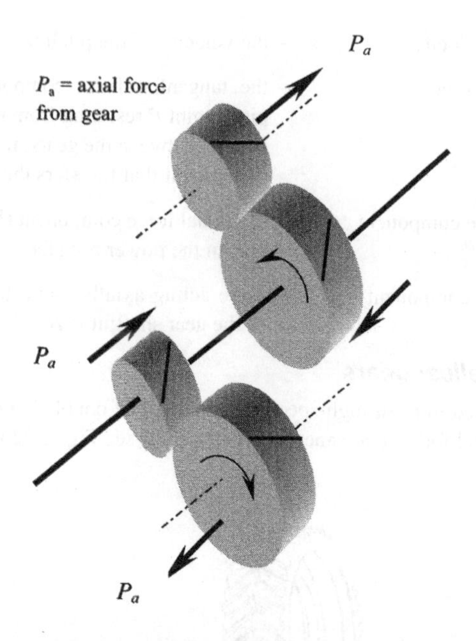

Fig. 5.23 Double helical gears: forces

Epicyclic gear sets

An epicyclic gear set consists of internal and external gears, assembled into a concentric set. They are used when a high speed or torque ratio has to be achieved in a compact physical space. Various arrangements are possible, depending on whether internal or external gears are used and which parts of the gear assembly are held stationary. Figure 5.24(a) shows a 'sun and planet' arrangement in which the planet gear rotates freely on its axle. Figure 5.24(b) shows a different arrangement in which the central 'sun' gear is replaced by a large-diameter internal ring-gear.

Figure 5.25 shows a typical physical layout of a basic epicyclic gearbox.

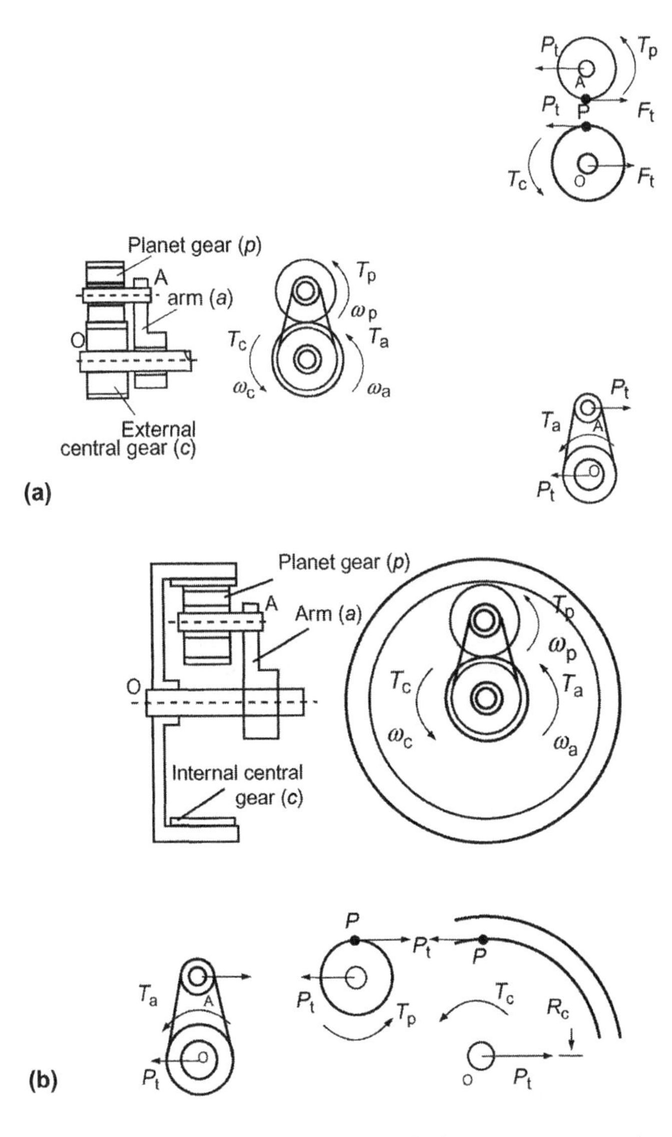

Fig. 5.24 Epicyclic – sun and planet arrangements

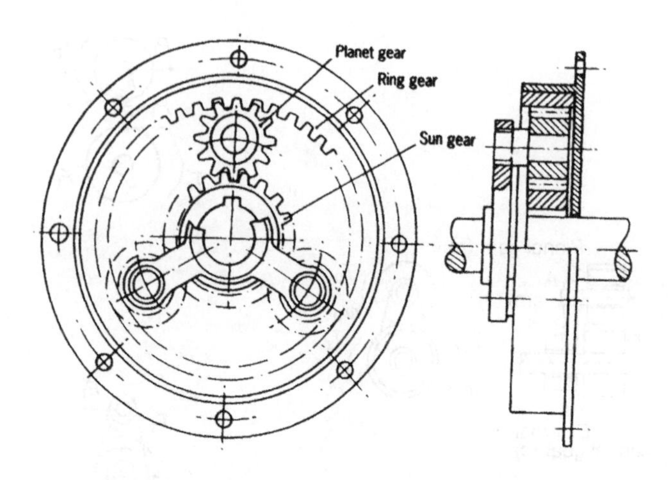

Fig. 5.25 Basic epicyclic gearbox layout

Tooth geometry and kinetics

Figure 5.24 shows torque and tangential force components, P_t, as they act on each of the gear components, represented as 'free body diagrams'. Note that each element has a single degree of 'kinetic' (torque) freedom, but two degrees of *kinematic* freedom. Key equations used in analyses are:

- $(\omega_c - \omega_a)z_c + (\omega_p - \omega_a)z_p = 0$
- $T_c/z_c = T_p/z_p = \dfrac{-T_a}{(z_c + z_p)}$

where z_c is a positive integer for an external central gear and a negative integer for an internal central gear.

ω_c = angular velocity of central gear
ω_p = angular velocity of planet gear
ω_a = angular velocity of connecting arm
T_c = torque on central gear shaft
T_p = torque on planet gear itself
T_a = torque on connecting arm

Gear selection

Table 5.7 shows basic information on gear selection of various rotating equipment applications.

Table 5.7 Gear selection – basic information

Type	*Features*	*Applications*	*Comments regarding precision*
Spur	• Parallel shafting • High speeds and loads • Highest efficiency	Applicable to all types of trains and a wide range of velocity ratios.	Simplest tooth elements offering maximum precision. Suitable for all gear meshes, except where very high speeds and loads or special features of other types, such as right angle drive, cannot be avoided.
Helical	• Parallel shafting • Very high speeds and loads • Efficiency slightly less than spur mesh	Most applicable to high speeds and loads.	Equivalent quality to spurs except for complication of helix angle. Suitable for all high-speed and high-load meshes. Axial thrust component must be accommodated.
Crossed-helical	• Skewed shafting • Point contact • Low speeds • Light loads	Relatively low velocity ratio; low speeds and light loads only. Any-angle skew shafts.	Not suitable for precision meshes. Point contact limits capacity and precision. Suitable for right-angle drives under light load. Good lubrication essential because of point contact and high sliding action.

Table 5.7 Cont.

Internal spur	• Parallel shafts • High speeds • High loads	Internal drives requiring high speeds and high loads; offers low sliding and high stress loading. Used in planetary gears to produce large reduction ratios.	Not suitable for precision meshes because of design, fabrication, and inspection limitations.
Bevel	• Intersecting shafts • High speeds • High loads	Suitable for 1:1 and higher velocity ratios and for right-angle meshes.	Suitable for right-angle drive, particularly low ratios. Complicated tooth form and fabrication limits achievement of precision.
Worm mesh	• Right-angle skew shafts • High velocity ratio • High speeds and loads • Low efficiency	High velocity ratio. Angular meshes. High loads.	Worm can be made to high precision, but the worm gear has inherent limitations. Suitable for average precision meshes. Best choice for combination high velocity ratio and right-angle drive. High sliding requires adequate lubrication.

Gear nomenclature

Gear standards refer to a large number of critical dimensions of the gear teeth. These are controlled by tight manufacturing tolerances.

Gear materials

Table 5.8 shows basic information on applications of gear materials.

Table 5.8 Gear materials – basic information

Material	*Features*	*Application*
Ferrous		
Cast irons	Low cost, good machining, high internal damping.	Large-size, moderate power rating commercial gears.
Cast steels	Low cost, high strength.	Power gears, medium ratings.
Plain-carbon steels	Good machining, heat treatable.	Power gears, medium ratings.
Alloy steels	Heat treatable, highest strength and durability.	Severest power requirements.
Stainless steels		
AISI 300 series	High corrosion resistance, non-magnetic, non-hardenable.	Extreme corrosion, low power ratings.
AISI 400 series	Hardenable, magnetic moderate stainless steel properties.	Low to medium power ratings, moderate corrosion.
Non-ferrous		
Aluminium alloys	Light weight, non-corrosive, excellent machinability.	Extremely light duty instrument gears.
Brass alloys	Low cost, non-corrosive, excellent machinability.	Low-cost commercial equipment.
Bronze alloys	Excellent machinability, low friction, and good compatibility with steel mates.	Mates for steel power gears.
Magnesium alloys	Extremely light weight, poor corrosion resistance.	Special lightweight, low-load uses.
Nickel alloys	Low coefficient of thermal expansion, poor machinability.	Special thermal cases.
Titanium alloys	High strength for moderate weight, corrosion resistant.	Special lightweight strength applications.
Die-cast alloys	Low cost, no precision, low strength.	High production, low quality, commercial.
Sintered powder alloys	Low cost, low quality, moderate strength.	High production, low quality commercial.

Table 5.8 Cont.

Non-metallic		
Delrin	Wear resistant, long life, low water absorption.	Long life, low noise, low loads.
Phenolic laminates	Quiet operation, highest strength plastic.	Medium loads, low noise.
Nylons	Low friction, no lubricant, high water absorption.	Long life, low noise, low loads.
Teflon (flurocarbon)	Low friction, no lubricant.	Special low friction.

Table 5.9 shows gear forces for various types of gear.

5.6 Seals

Seals are used either to provide a seal between two working fluids or to prevent leakage of a working fluid to the atmosphere past a rotating shaft. There are several types.

Bellows seal

This uses a flexible bellows to provide pressure and absorb misalignment (see Fig. 5.26).

Labyrinth gland

This consists of a series of restrictions formed by projections on the shaft and/or casing (see Fig. 5.27). The pressure of the steam or gas is broken down by expansion at each restriction. There is no physical contact between the fixed and moving parts.

Mechanical seals

Mechanical seals are used either to provide a seal between two working fluids or to prevent leakage of a working fluid to the atmosphere past a rotating shaft. This rotary motion is a feature of mechanical seals. Other types of seal are used for reciprocating shafts, or when all the components are stationary. Figure 5.28 shows a typical mechanical seal and Fig. 5.29 a specific design with its component pieces. They can work with a variety of fluids and, in the extreme, can seal against pressures of up to 500 bar, and have sliding speeds of more than 20 m/s. The core parts of the seal are the rotating 'floating' seal ring and the stationary seat. Both are made of wear-resistant materials and the floating ring is kept under axial force from a

Table 5.9 Formulae for gear forces

		Tangential force P_t	***Radial force P_r***	***Axial force P_a***
Cylindrical gears	Spur	$\frac{2M_t}{d_1}$	$P_t \tan \propto$	
	Helical	$\frac{2M_t}{d_1}$	$P_t \frac{\tan \propto}{\cos\beta}$	$P_t \tan \beta$
	Herringbone	$\frac{2M_t}{d_1}$	$P_t \frac{\tan \propto}{\cos\beta}$	
Bevel gears	Straight	$(P_t)_{AV} = \frac{2M_t}{d_{AV}} = \frac{2M_t}{d_1(1-0.5\ b/R)}$	$(P_t)_{AV} \tan \propto \cos\delta$	$(P_t)_{AV} \tan \propto \sin\delta$
	Spiral	$(P_t)_{AV} = \frac{2M_t}{d_{AV}} = \frac{2M_t}{d_1(1-0.5\ b/R)}$	$\frac{(P_t)_{AV}}{\cos\beta_{AV}}[\tan \propto \cos\delta \pm \sin\beta_{AV} \sin\delta]$	$\frac{(P_t)_{AV}}{\cos\beta_{AV}}[\tan \propto \sin\delta \pm \sin\beta_{AV} \cos\delta]$
Worm gears	Wheel	$\frac{2M_t}{d_2}$	$P_t \tan \propto$	$P_t \tan(\gamma + p)$
	Worm	$P_1 \tan(\gamma + p)$	$P_t \tan \propto$	$\frac{2M_t}{d_2}$

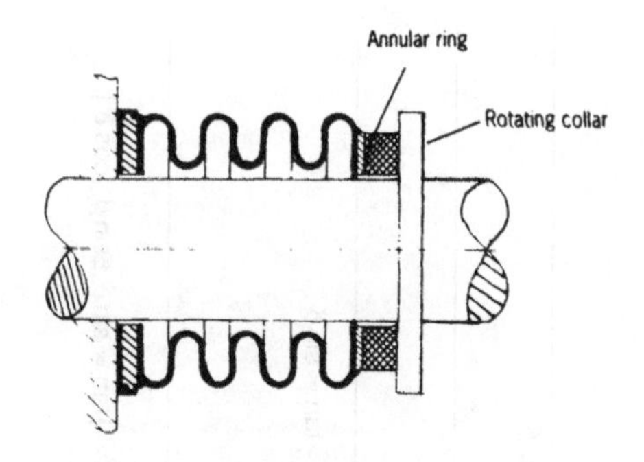

Fig. 5.26 Bellows seal

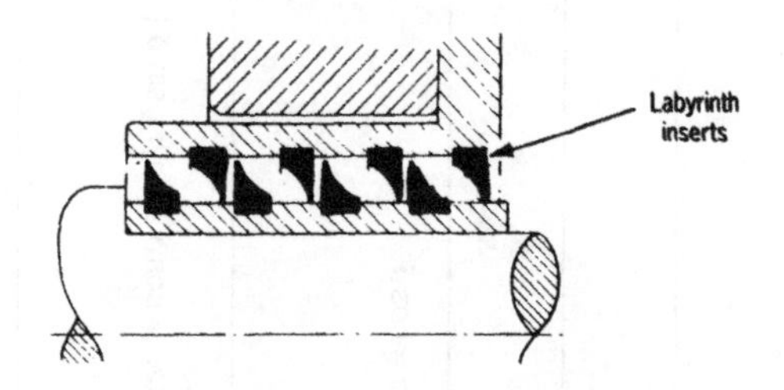

Fig. 5.27 Labyrinth gland

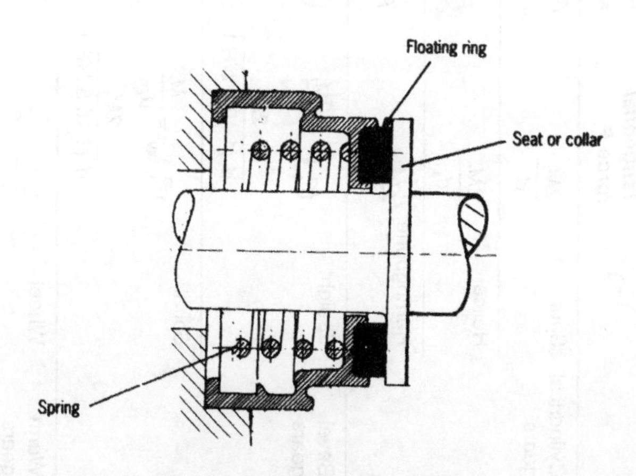

Fig. 5.28 Mechanical seal

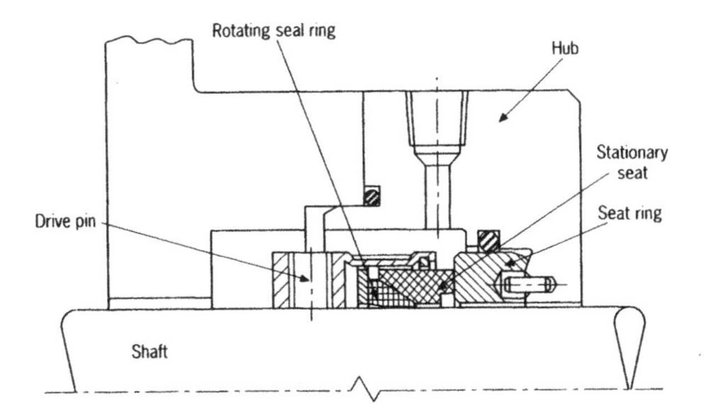

Fig. 5.29 Mechanical seal

spring or bellows to force it into contact with the seat face. This is the most common type and is termed a 'face seal'. It is found in common use in many engineering applications: vehicle water pumps and automatic transmission gearboxes, washing machines and dishwashers, as well as more traditional industrial use on most types of process pumps. Materials of construction are quite varied, depending on the characteristics of the process fluid.

Table 5.10 Mechanical seal notation

A	Seal area ratio
A_h	Hydraulic area
A_i	Interface area
b	Seal interface width
C	A 'shape factor'
d	Internal diameter of seal interface
d_h	Recess diameter
D	External diameter of seal interface
E	Young's modulus of seal ring
M	Moment arm
P	Sealed fluid pressure (external)
p_i	Sealed fluid pressure (internal)
r	Internal radius of a ring
r_m	Mean radius of seal interface
r_p	Torque arm radius
R	External radius of a ring
s	Deflection of seal ring
W_f	Friction force
W_h	Net hydraulic force
W_o	Opening force
W_s	Spring force
σ_b	Compressive stress at seal ring bore
σ_z	Tensile stress at seal ring bore
ϕ	Angular distortion of ring

Table 5.10 shows typical mechanical seal notation.

Mechanical seals are mass-produced items manufactured in a large range of sizes. Special designs are required for use with aggressive process fluids such as acids, alkalis, and slurries. The design of mechanical seals is specialized and has developed iteratively over many years using a mixture of engineering disciplines such as:

- *Thermodynamics*. Heat transfer in the seal components and the fluid film must be considered.
- *Fluid mechanics*. Hydrodynamic, boundary, and static lubrication conditions exist in various areas of the sealing face and associated parts of the seal assembly. Laminar flow calculations govern the leakage path between the floating seal ring and stationary seat. Static fluid pressure considerations are used to determine the additional axial 'sealing force' generated by the process fluid.

- *Deformable body mechanics*. Deformation of the seal ring in use is an important design parameter. This is calculated using classical 'hollow cylinder' assumptions with 'open-end' boundary conditions. Local deformations of the seal and seat faces are important (see Fig. 5.30).
- *Surface mechanics*. Surface characteristics, particularly roughness profile, of the contacting faces affect leakage, friction, and wear. This plays an important part in the tribology (the study of moving surfaces in contact) of the sealing faces.
- *Materials technology*. Material properties also play a part in the tribology of mechanical seals. The compatibility of the seal faces, wear resistance, and friction characteristics are influenced by the choice of materials – many seals use very specialized wear-resistant materials such as plastics or ceramics.

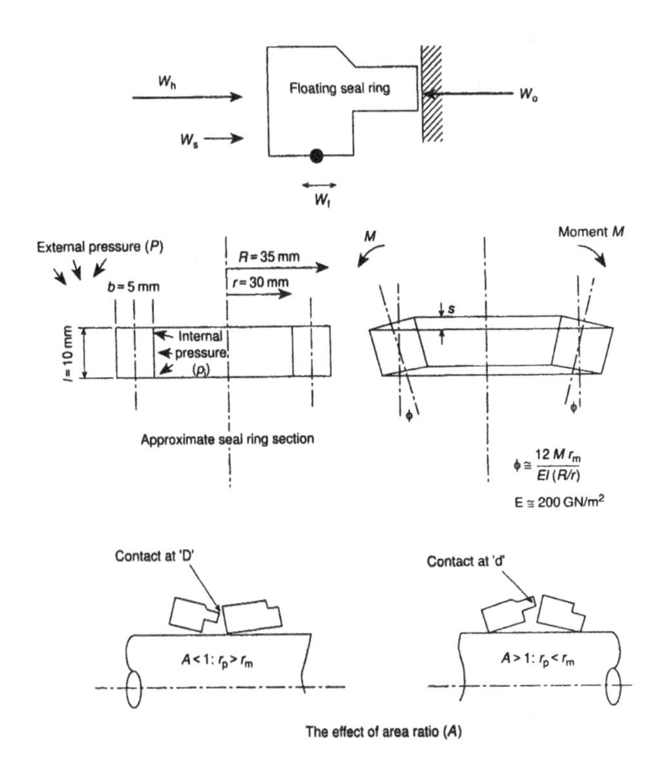

Fig. 5.30 Mechanical seal ring behaviour

Seal area ratio

In practice, most mechanical seals do not rely only on the force of the spring to keep the seal faces in contact (termed 'closure'). Closure is mainly achieved by the net hydraulic fluid pressure acting on the seal floating ring. This net hydraulic pressure is a function of the differential areas of the floating ring – hence the closure force increases as the sealed fluid pressure increases, and the spring actually plays little part. Figure 5.30 shows how the closure force is made up of four components. The net hydraulic force W_h comes from the sealed fluid at pressure p_i. This is joined by the spring force W_s, the 'opening' force W_o caused by the seal interface fluid pressure, and the friction force W_f caused by the frictional resistance of the static seal. Opening force W_o is normally calculated using the assumption that fluid pressure varies in a linear way across the radial seal face. The frictional resistance force W_f is just about indeterminable and is often ignored. Note the locations of the static 'O' ring seals; they are an essential part of the seal assembly, to eliminate static leakage paths.

Seal ring dimensions

Mechanical seal design includes static calculations on the seal rings, which must have a sufficient factor of safety to avoid bursting. A valid assumption used is that face seal rings behave as hollow cylinders with open ends. Hence, for a hollow cylinder with internal radius r, external radius R, subject to internal pressure p_i and external pressure P it can be shown that

Maximum stress (Lamé)

$$\sigma = \frac{p_i(R^2 + r^2) - 2PR^2}{R^2 - r^2}$$

For internal pressure (i.e. when $P = 0$) then maximum tensile stress, σ_z, at the ring bore is given by

$$\sigma_z = \frac{p_i(R^2 + r^2)}{R^2 - r^2}$$

or, for external pressure (i.e. where $p_i = 0$), then maximum compressive stress σ_b at the bore is given by

$$\sigma_b = \frac{2P}{1 - \left(r/R\right)^2}$$

These equations only apply, strictly, to seal rings of plain rectangular cross-section. They can be used as an order-of-magnitude check, however; then significant factors of safety are included to allow for any uncertainties.

Seal ring deflections

A further important design criterion is the twisting moment which occurs in the floating seal ring. This can cause deflection (distortion) of the seal ring surface. From Figs 5.29 and 5.30:

D = external diameter of seal interface
d = internal diameter of seal interface
d_h = recess diameter
r_m = mean radius = (D + d)/4
r_p = torque arm radius
b = seal interface width

The moment arm $M = \bar{P}(r_p - r_m)$
where

$$\bar{P} = AP\ b$$

and

$$r_p = \frac{D + d_h}{4}$$

As the area ratio A tends towards being greater or smaller than unity then the lever distance $(r_p - r_m)$ gets larger, hence increasing the moment M. This moment produces angular distortion of the floating seal ring. Referring again to Fig. 5.30 the angular distortion ϕ is given by

$$\phi = \frac{12 M r_m}{E l^3 (R/r)}$$

This results in a physical deflection s given by

$$s = \phi b C$$

where C is a shape factor (near unity) related to the section of the floating ring.

The result of this is that the floating ring will contact at one end of its surface. As a general rule for rings under external pressure:

- if $A < 1$, the floating ring twists 'inwards' towards the shaft and hence contacts at its outer 'D' edge.
- if $A > 1$, the floating ring twists 'outwards' away from the shaft and hence contacts at its inner 'd' edge.

It is essential, therefore, when considering seal design, to calculate any likely twist of the floating seal rings to ensure that this is not so excessive that it reduces significantly the seal interface contact area. A maximum distortion s of 15 microns is normally used as a rule of thumb.

Friction considerations

The floating seal ring/seat interface is the main area of a mechanical seal in which friction is an issue. The whole purpose of this interface is to provide the main face sealing surface of the assembly with only a controlled degree of leakage, hence some friction at this face is inevitable. If it becomes too high, too much heat will be generated, which may cause excessive distortion of the components and eventual seizure. Given that rotational speed of the shaft is difficult to change (it is decided by the process requirements of the pump or machine) careful choice of the ring materials and their respective surface finish is the best way of keeping friction under control.

Surface finish is defined using the parameter R_a measured in microns. This is the average distance between the centreline of a surface's undulations and the extremes of the peaks and the troughs. It is sometimes referred to as the centre line average (CLA); see Chapter 11, Fig. 11.9. If the surface is too rough, the effective contact area of the interface will be reduced, resulting in a significant increase in seal loading for a constant closure force W. This can cause lubrication film breakdown and seizure. Experience shows that the lubrication regime existing between the interface surfaces is rarely completely hydrodynamic; boundary lubrication conditions provide a better assumption and these are prone to breakdown if specific contact loading is too high. Conversely, a surface that is too smooth is less able to 'hold' the lubricant film so, while the effective contact area of the interface is increased with smooth surfaces, there may be negative effects on the stability of the lubrication regime. In practice it has been found that a surface roughness of 0.1 ± 0.025 μm R_a (both contacting rings) gives the best results.

Assembly considerations

Many mechanical seals fail in the early stages of their life because of inaccurate assembly. This is normally due to one of two reasons.

- *Axial misalignment*. This is displacement of the seal end housing, and/or the static seat sealing ring so that its centreline lies at an angle to that of the rotating parts. This results in almost instantaneous wear and failure after only a few running hours. Misalignment can be prevented by incorporating features that give positive location of the seal end housing.
- *Poor concentricity*. This is mainly a fault with the positioning of the fixed seal seating ring centreline, i.e. it is not concentric with that of the shaft. It can be almost eliminated by specifying the manufactured concentricity level of the components and then providing a positive concentric location for the seal seating ring in the end housing. Note that this must still incorporate the 'O' ring, which prevents fluid leakage.

To avoid sharp changes in section of the rotating shaft, while still providing an abutment face for the closure spring, some seal designs incorporate a shaft sleeve. This fits concentrically over the shaft for slightly more than the length of the spring. The static 'O' ring rubber seals are often a problem during assembly as they can be chafed by the sharp edges of their slots. This causes the 'O' ring to lose its 100 per cent sealing capability.

5.7 Cam mechanisms

A cam and follower combination are designed to produce a specific form of output motion. The motion is generally represented on a displacement/time (or lift/angle) curve. The follower may have a knife-edge, roller, or flat profile.

Constant velocity cam

This produces a constant follower speed and is only suitable for simple applications (see Fig. 5.31).

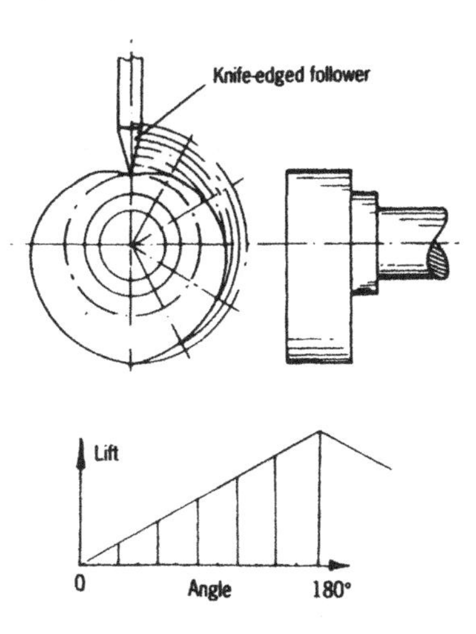

Fig. 5.31 Constant velocity cam

Uniform acceleration cam

The displacement curve is a second-order function giving a uniformly increasing/decreasing gradient (velocity) and constant d^2x/dt^2 (acceleration). See Fig. 5.32.

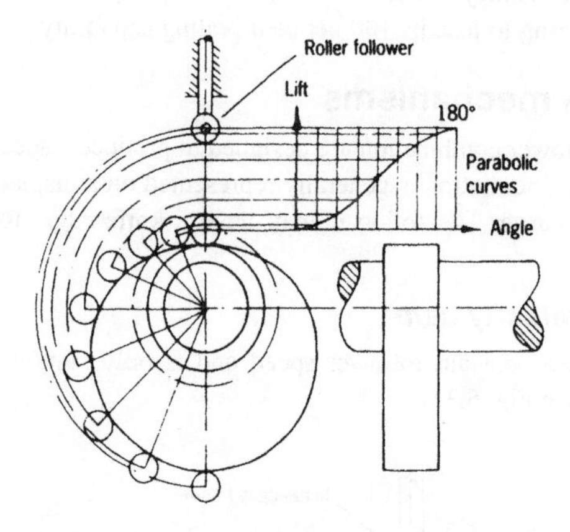

Fig. 5.32 Uniform acceleration cam

Simple harmonic motion cam

A simple eccentric circle cam with a flat follower produces simple harmonic motion (see Fig. 5.33).

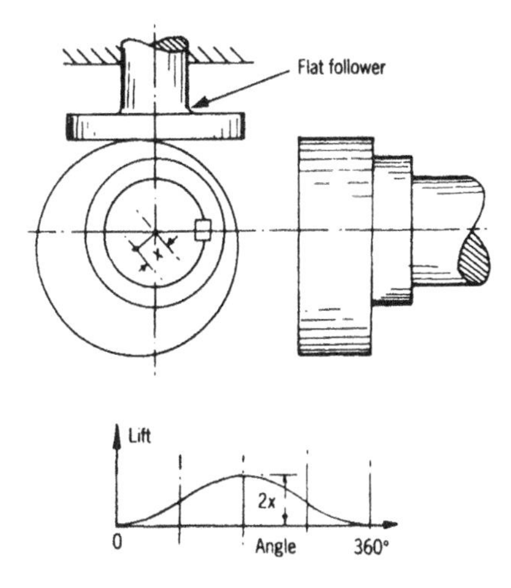

Fig. 5.33 Simple harmonic motion cam

The motion follows the general harmonic motion equation

$$d^2x/dt^2 = -\omega^2 x$$

where

x = displacement
ω = angular velocity
T = periodic time
$dx/dt = -\omega a \sin \omega t$
$T = 2\pi/\omega$

5.8 Belt drives

Types

The most common types of belt drive are flat, 'V', wedge, and ribbed (see Fig. 5.34).

- *Flat belts* are weak and break easily – their use is limited to a few low-torque high-speed applications.
- *'V' belts* provide a stronger and more compact drive than a flat belt and comprise cord tensile strands embedded in the matrix of the belt material, in the region of the pitchline.

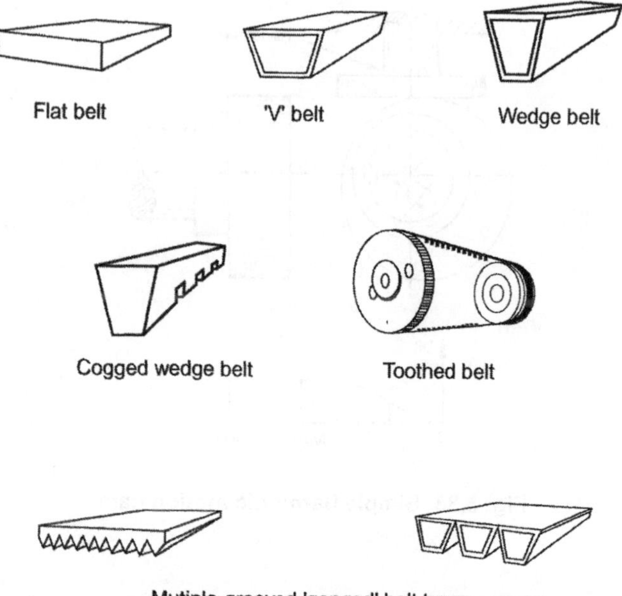

Fig. 5.34 Belt drive types

- A variant of the 'V' belt is the lighter and narrower *wedge belt*. The lighter weight means that centrifugal forces (which reduce the driving friction of the belt in the pulley grooves) are lower, hence the belt provides better drive at higher speeds than a plain 'V' belt.
- A further variant is the *cogged wedge belt* – this uses transverse slots or recesses on the underside of the belt to enable the belt to bend more easily around smaller diameter pulleys. The slots play little or no effective part in the driving action.
- The most advanced type of belt drive is the *toothed belt*, which gives a positive drive using gear-like teeth.

For higher power transmission requirements, multiple or ganged belts are used on multi-grooved pulleys. All belts are manufactured in a range of standard cross-sectional sizes carrying various designations.

'V' belt geometry

The main use for 'V' belts is in short-centre drives with speeds of around 20 m/s. Figure 5.35 shows an indicative range of size/power characteristics and Fig. 5.36 the basic geometry of the drive.

5.9 Clutches

Clutches are used to enable connection and disconnection of driver and driven shafts.

Jaw clutch

One half of the assembly slides on a splined shaft. It is moved by a lever mechanism into mesh with the fixed half on the other shaft. The clutch can only be engaged when both shafts are stationary. Used for crude and slow-moving machines such as crushers (see Fig. 5.37).

Cone clutch

The mating surfaces are conical and normally lined with friction material. The clutch can be engaged or disengaged when the shafts are in motion. Used for simple pump drives and heavy duty materials handling equipment (see Fig. 5.38).

Multi-plate disc clutch

Multiple friction-lined discs are interleaved with steel pressure plates. A lever or hydraulic mechanism compresses the plate stack together. Universal use in motor vehicles with manual transmission (see Fig. 5.39).

Fluid couplings

Radial-vaned impellers run in a fluid-filled chamber. The fluid friction transfers the drive between the two impellers. Used in automatic transmission motor vehicles and for larger equipment such as radial fans and compressors (see Fig. 5.40).

The key design criterion of any type of friction clutch is the axial force required in order to prevent slipping. A general formula is used, based on the assumption of uniform pressure over the contact area (see Fig. 5.41).

$$\text{Force } F = \frac{3T\,(r_2^2 - r_1^2)}{2f(r_2^3 - r_1^3)}$$

T = torque
f = coefficient of friction

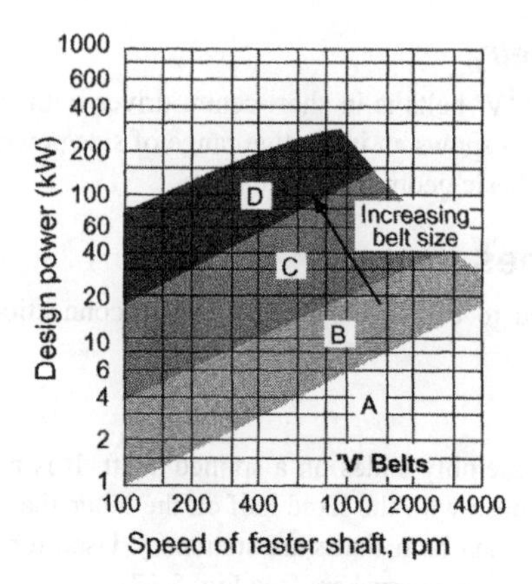

A, B, C, D are indicative ranges of belts based on cross-sectional area

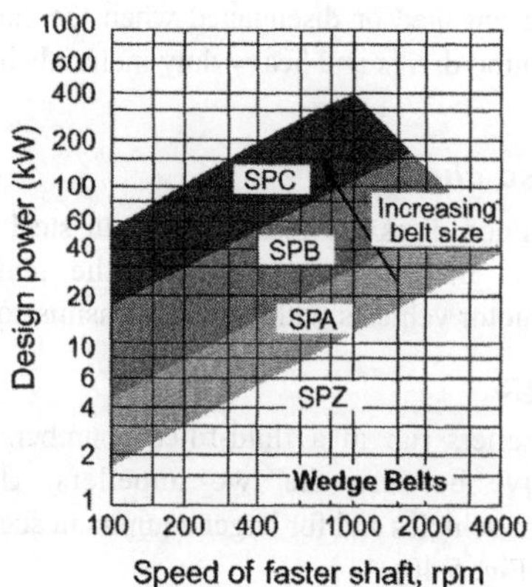

SPZ, SPA, SPB, SPC are indicative ranges of belts based on cross-sectional area

Fig. 5.35 'V'/wedge belts – indicative size:power characteristics

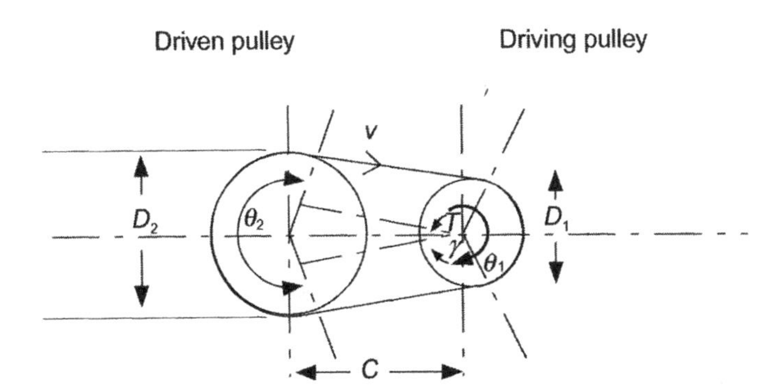

C = Centre distance
v = Belt speed
R = Drive ratio $\equiv D_2/D_1$
$\theta_1 = \theta_2$ = wrap angle on each pulley = $\pi - 2\gamma$
β = Belt grove 'semi' angle $\cong$ 16–20°

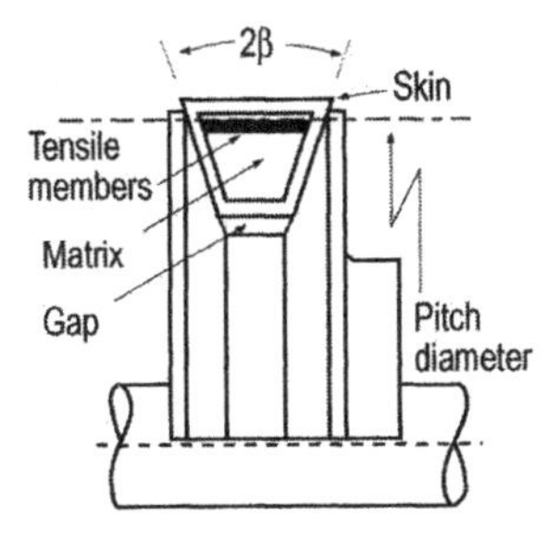

Fig. 5.36 'V'/wedge belts – basic geometry

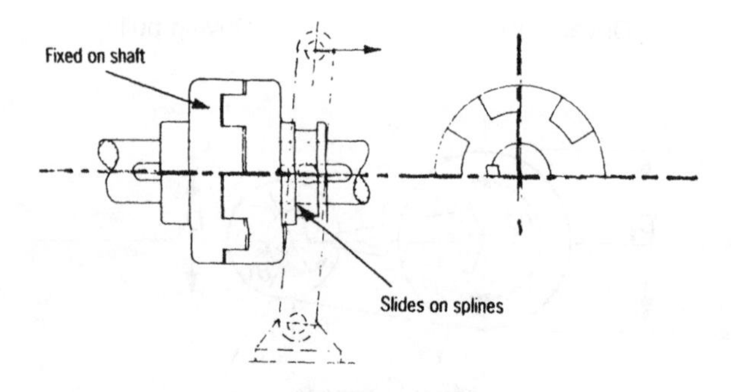

Fig. 5.37 Jaw clutch

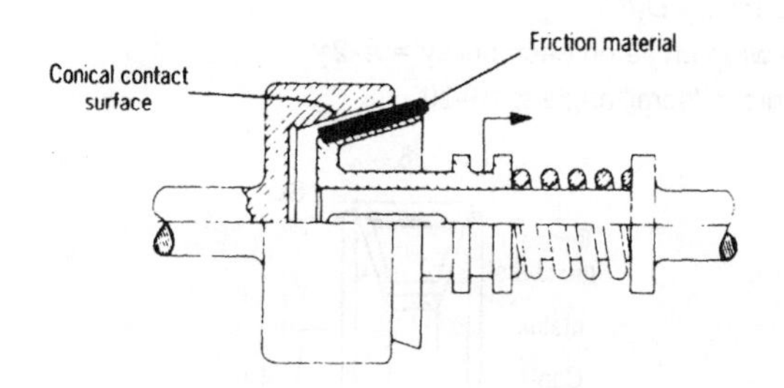

Fig. 5.38 Cone clutch

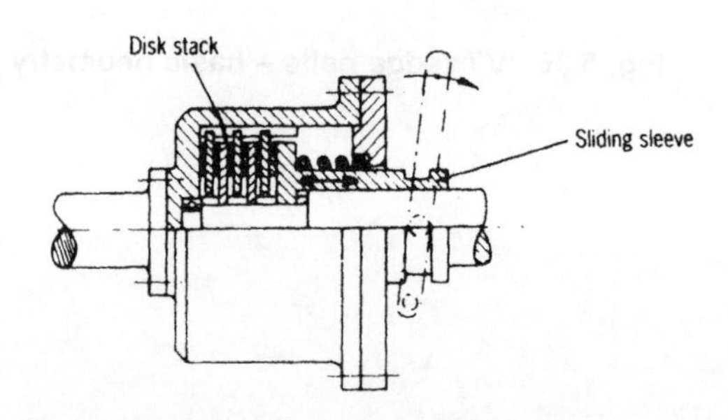

Fig. 5.39 Multi-plate disc clutch

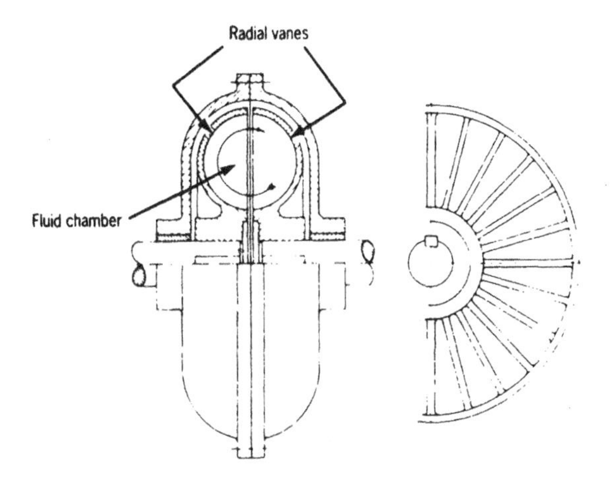

Fig. 5.40 Fluid coupling

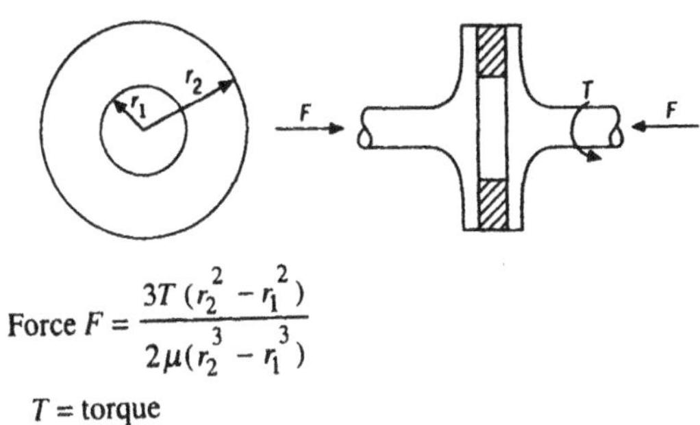

$$\text{Force } F = \frac{3T\,(r_2^2 - r_1^2)}{2\mu(r_2^3 - r_1^3)}$$

T = torque
μ = coefficient of friction

Fig. 5.41 Clutch friction

5.10 Brakes

Brake types

Brakes are used to decelerate a rotating component or system of components by absorbing power from it. Most types use simple sliding friction. Figure 5.42 shows some basic models.

- The simple *band brake* comprises a flexible band bearing on the circumference of a drum; these are used on simple winches.
- The *external shoe brake* has external shoes with friction linings, rigidly connected to pivoted posts. The brake is operated by a linkage, which provides an actuation force, pulling the brake shoes into contact with the drum.
- *Internal drum brakes*, used on older designs of motor vehicle, operate by the friction-lined brake shoes being pushed into contact with the internal surface of a brake drum by a single cam (leading/trailing leading shoe type) or twin hydraulic cylinders (twin leading shoe type).
- *Hydraulic disc brakes*, as found on most road vehicles, aircraft, etc. and many industrial applications comprise twin opposing hydraulic pistons faced with pads of friction material. The pads are forced into contact with the disc by hydraulic pressure, exerting forces normal to the disc which transfer into tangential friction forces, thereby applying a deceleration force to the disc.

5.11 Pulley mechanisms

Pulley mechanisms can generally be divided into either 'simple' or 'differential' types.

Simple pulleys

These have a continuous rope loop wrapped around the pulley sheave. The key design criterion is the velocity ratio (see Fig. 5.43).

Velocity ratio, VR = the number of rope cross-sections supporting the load.

Differential pulleys

These are used to lift very heavy loads and consist of twin pulleys 'ganged' together on a single shaft (see Fig. 5.44).

$$VR = \frac{2\pi R}{\pi(R-r)} = \frac{2R}{R-r}$$

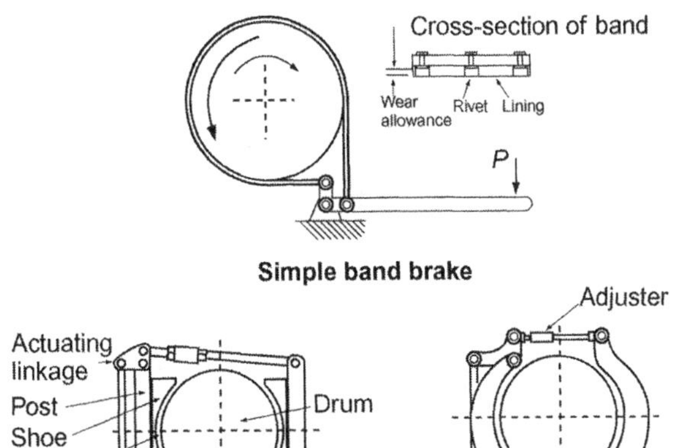

Simple band brake

Two designs of external shoe drum brake

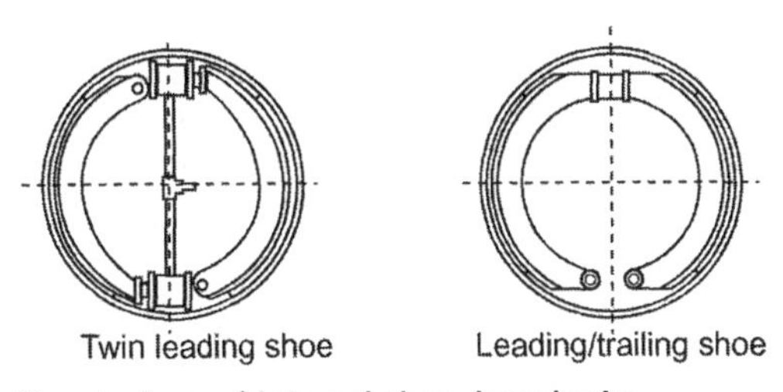

Two designs of internal shoe drum brake

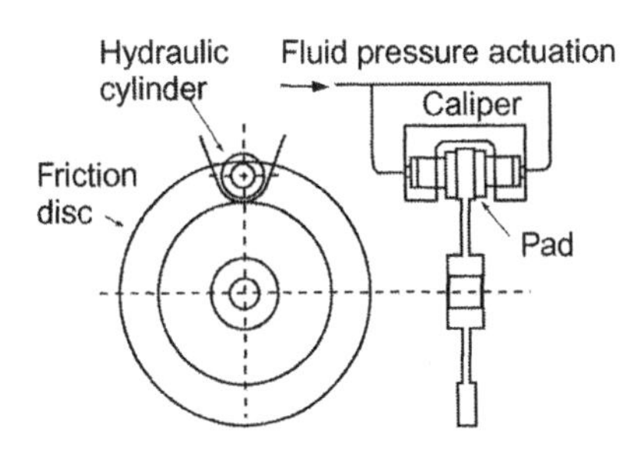

Hydraulic disk brake

Fig. 5.42 Brakes – basic types

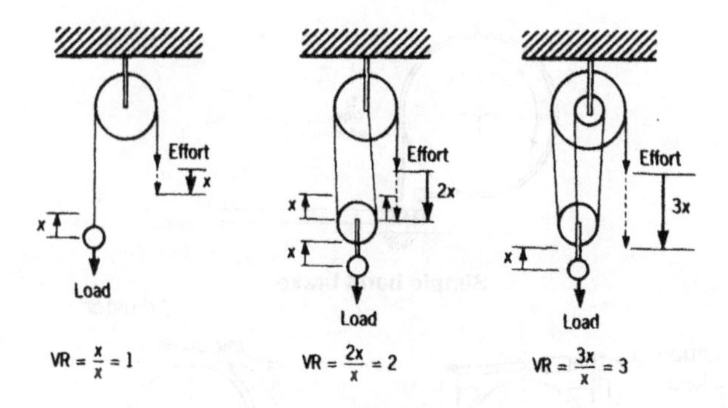

Fig. 5.43 Simple pulleys

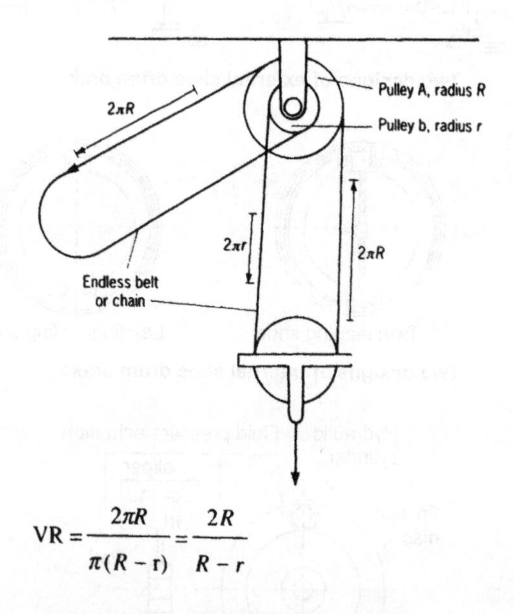

$$\mathrm{VR} = \frac{2\pi R}{\pi(R - r)} = \frac{2R}{R - r}$$

Fig. 5.44 Differential pulleys

5.12 Useful references and standards

Standards: bearings

1. BS 5983 Part 6: 1983 *Metric spherical plain bearings – glossary of terms.*
2. BS 5512: 1991 *Method of calculating load ratings and rating life of roller bearings.* This is an equivalent standard to ISO 281.
3. BS 292: Part 1: 1987 *Specification for dimensions of ball, cylindrical and spherical roller bearings (metric series).*
4. BS 5645: 1987 *Glossary of terms for roller bearings.* Equivalent to ISO 76.
5. BS 5989: Part 1: 1995 *Specification for dimensions of thrust bearings.* Equivalent to ISO 104: 1994.
6. BS 6107 (various parts). *Rolling bearings – tolerances.*
7. BS ISO 5593: 1997 *Rolling bearings – vocabulary.*
8. ABMA A24.2: 1995 *Bearings of ball, thrust and cylindrical roller types – inch design.*
9. ABMA A20: 1985 *Bearings of ball, radial, cylindrical roller, and spherical roller types – metric design.*
10. ISO 8443: 1989 *Rolling bearings.*
11. ANSI/ABMA/ISO 5597: 1997 *Rolling bearings – vocabulary.*

Bearing websites

1. Anti-Friction Bearing Manufacturers Association Inc: www.afbma.org
2. www.skf.se/products/index.htm
3. www.nsk-ltd.co.jp

Standards: gears

1. ISO 1328: 1975 *Parallel involute gears – ISO system of accuracy.*
2. ANSI/AGMA 2000-A88 1994 *Gear classification and inspection handbook.*
3. ANSI/AGMA 6002: B93 1999 *Design guide for vehicle spur and helical gears.*
4. ANSI/AGMA 6019-E89: 1989 *Gear-motors using spur, helical herringbone straight bevel, and spiral bevel gears.*
5. BS 436: Part 1: 1987 *Basic rack form, pitches, and accuracy.*
6. API 613: 1998 *Special-purpose gear units for refinery service.*
7. DIN 3990: 1982 *Calculation of load capacity of cylindrical gears.*

Gear websites

1. www.agma.org
2. www.Reliance.co.uk
3. www.flender.com

Standards: seals

Mechanical seals are complex items and manufacturers' in-house (confidential) standards tend to predominate. Some useful related standards are:

1. MIL S-52506D: 1992 *Mechanical seals for general purpose use.*
2. KS B1566: 1997 *Mechanical seals.*
3. JIS (Japan) B2405: 1991 *Mechanical seals – general requirements.*
4. BS 6241: 1982 *Specification of housings for hydraulic seals for reciprocating applications.* This is a similar standard to ISO 6547.

Seal websites

1. www.flexibox.com
2. www.garlock-inc.com

Standards: couplings

1. BS 6613: 1991 *Methods of specifying characteristics of resilient shaft couplings.* Equivalent to ISO 4863.
2. BS 3170: 1991 *Specification for flexible couplings for power transmission.*
3. BS 5304: 1988 *Code of practice for safety of machinery.* This includes details on the guarding of shaft couplings.
4. API 617: 1990 *Special-purpose couplings for refining service.*
5. AGMA 515 *Balance classification for flexible couplings.*
6. KS B1555 *Rubber shaft couplings.*
7. KS B1553 *Gear type shaft couplings.*
8. ISO 10441: 1999 *Flexible couplings for mechanical power transmission.*
9. ANSI/AGMA 9003-A91 (R1999) *Flexible couplings – keyless fits.*

Standards: clutches

1. BS 3092: 1988 *Specification for main friction clutches for internal combustion engines.*
2. SAE J2408: 1984 *Clutch requirements for truck and bus engines.*

Standards: pulleys

1. SAE J636: 1992 *'V' belts and pulleys.*
2. BS 3876: Part 2: 1990 *Specification for vertical spindle pulleys, mountings and assemblies.*

Standards: belt drives

1. BS 7620: 1993 *Specification for industrial belt drives – dimensions of pulleys and 'V' ribbed belts of PH, PJ, PK, PL, and PM profiles*. Similar to ISO 9982.
2. BS 4548: 1987 *Specification for synchronous belt drives for industrial applications*. Similar to ISO 5294.
3. BS AU 150b: 1990 *Specification for automotive 'V' belts and pulleys.*
4. BS AU 218: 1987 *Specification for automotive synchronous belt drives.*
5. SAE J637: 1998 *'V' belt drives.*
6. RMA-IP20: 1987 *Specification for drives – 'V' belt and sheaves.*
7. ISO 22: 1991 *Belt drives – flat transmission belts.*
8. ISO 4184: 1992 *Belt drives – classical and narrow 'V' belts.*

CHAPTER 6

Fluid Mechanics

6.1 Basic properties

Basic relationships

Fluids are classified into liquids, which are virtually incompressible, and gases, which are compressible. A fluid consists of a collection of molecules in constant motion: a liquid adopts the shape of a vessel containing it, while a gas expands to fill any container in which it is placed. Some basic fluid relationships are given in Table 6.1.

Table 6.1 Basic fluid relationships

Density, ρ	Mass per unit volume. Units kg/m^3 (lb/in^3)
Specific gravity, s	Ratio of density to that of water, i.e. $s = \rho/\rho_{water}$
Specific volume, v	Reciprocal of density, i.e. $v = 1/\rho$. Units m^3/kg (in^3/lb)
Dynamic viscosity, μ	A force per unit area or shear stress of a fluid. Units Ns/m^2 (lbf.s/ft^2)
Kinematic viscosity, ν	A ratio of dynamic viscosity to density, i.e. $\nu = \mu/\rho$. Units m^2/s (ft^2/s)

Perfect gas

A perfect (or 'ideal') gas is one that follows Boyle's/Charles's law

$$pv = RT$$

where

p = pressure of the gas

v = specific volume

T = absolute temperature

R = the universal gas constant

Although no actual gases follow this law totally, the behaviour of most gases at temperatures well above their liquification temperature will approximate to it and so they can be considered as a perfect gas.

Changes of state

When a perfect gas changes state its behaviour approximates to

$$pv^n = \text{constant}$$

where n is known as the polytropic exponent.

The four main changes of state relevant to rotating equipment are: isothermal, adiabatic, polytropic, and isobaric.

Compressibility

The extent to which a fluid can be compressed in volume is expressed using the compressibility coefficient β.

$$\beta = \frac{\Delta v / v}{\Delta p} = \frac{1}{K}$$

where

Δv = change in volume

v = initial volume

Δp = change in pressure

K = bulk modulus

Also

$$K = \rho \frac{\Delta p}{\Delta \rho} = \rho \frac{dp}{d\rho}$$

and

$$a = \sqrt{\left(\frac{dp}{d\rho}\right)} = \sqrt{\left(\frac{K}{\rho}\right)}$$

where

a = the velocity of propagation of a pressure wave in the fluid.

Fluid statics

Fluid statics is the study of fluids that are at rest (i.e. not flowing) relative to the vessel containing them. Pressure has four important characteristics:

- pressure applied to a fluid in a closed vessel (such as a hydraulic ram) is transmitted to all parts of the closed vessel at the same value (Pascal's law);
- the magnitude of pressure force acting at any point in a static fluid is the same, irrespective of direction;
- pressure force always acts perpendicular to the boundary containing it;
- the pressure 'inside' a liquid increases in proportion to its depth.

Other important static pressure equations are:

- absolute pressure = gauge pressure + atmospheric pressure
- pressure p at depth h in a liquid is given by
  $p = \rho g h$
- a general equation for a fluid at rest is

$$pdA - \left(p + \frac{\mathrm{d}p}{\mathrm{d}z}\right)\mathrm{d}A - \rho g \ \mathrm{d}A \ \mathrm{d}z = 0$$

This relates to an infinitesimal vertical cylinder of fluid.

6.2 Flow equations

Fluid flow in rotating equipment may be one-dimensional (1-D), two-dimensional (2-D), or three-dimensional (3-D) depending on the way that the flow is constrained.

One-dimensional flow

One-dimensional flow has a single direction co-ordinate x and a velocity in that direction of v. Flow in a pipe or tube is generally considered one-dimensional. The equations for 1-D flow are derived by considering flow along a straight stream tube (see Fig. 6.1). Table 6.2 shows the principles, and their resulting equations.

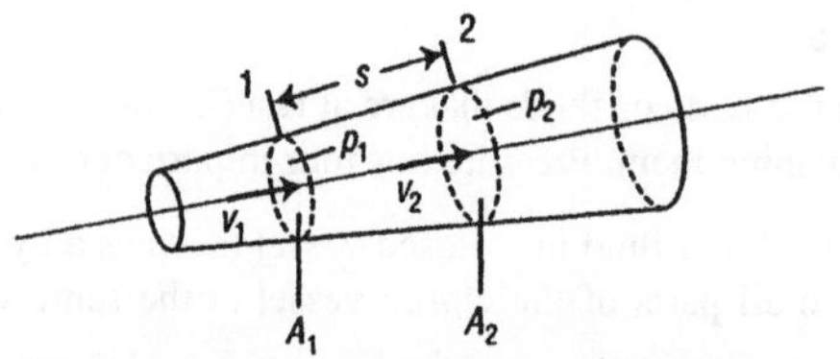

The stream tube for conservation of mass

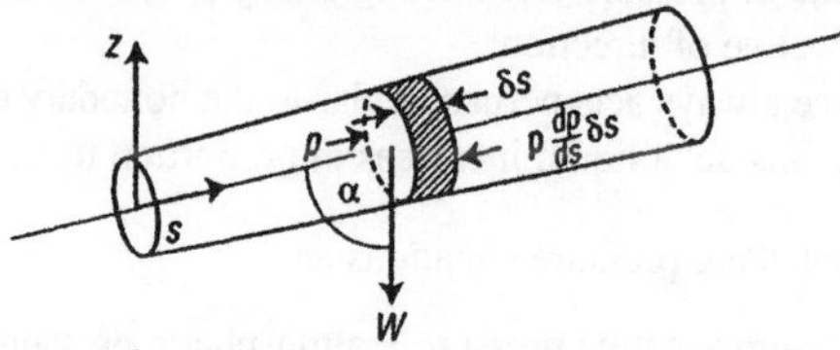

The stream tube and element for the momentum equation

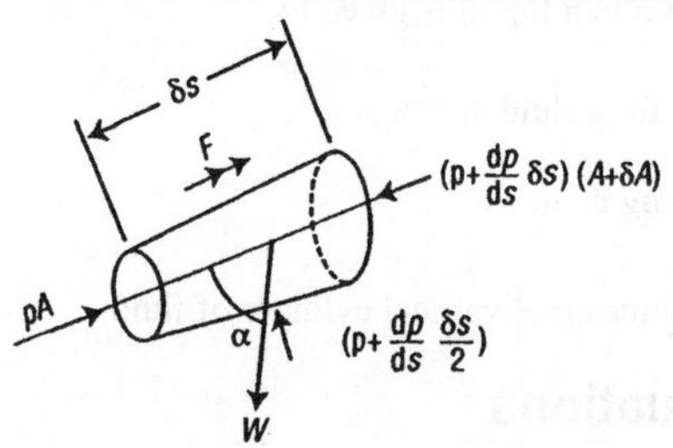

The forces on the element

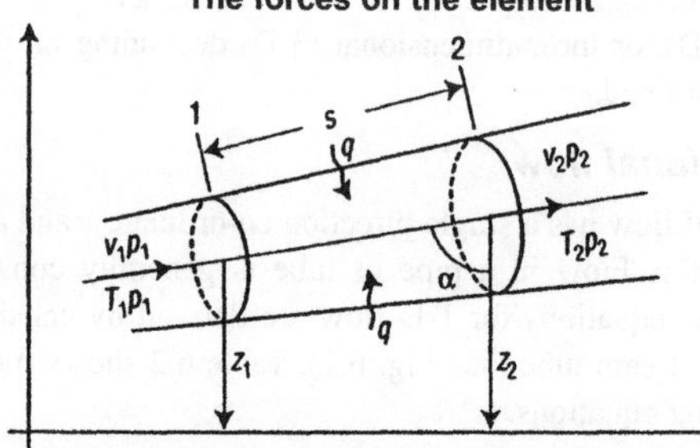

Control volume for the energy equation

Fig. 6.1 One-dimensional flow

Table 6.2 Fluid principles

Law	*Basis*	*Resulting equations*
Conservation of mass	Matter (in a stream tube or anywhere else) cannot be created or destroyed.	ρvA = constant
Conservation of momentum	The rate of change of momentum in a given direction = algebraic sum of the forces acting in that direction (Newton's second law of motion).	$\int\sqrt{\left(\frac{dp}{\rho}\right)} + \frac{1}{2}v^2 + gz$ = constant This is Bernoulli's equation
Conservation of energy	Energy, heat and work are convertible into each other and are in balance in a steadily operating system.	$c_pT + \frac{v^2}{2}$ = constant for an adiabatic (no heat transferred) flow system
Equation of state	Perfect gas state $p/\rho T = R$ and the first law of thermodynamics	$p = k\rho^\gamma$ k = constant γ = ratio of specific heats c_p/c_v

Two-dimensional flow

Two-dimensional flow (as in the space between two parallel flat plates) is that in which all velocities are parallel to a given plane. Either rectangular (x,y) or polar (r,θ) co-ordinates may be used to describe the characteristics of 2-D flow. Table 6.3 and Fig. 6.2 show the fundamental equations.

Table 6.3 Two-dimensional flow – fundamental equations

Basis	*The equation*	*Explanation*
Laplace's equation	$\frac{\partial^2 \phi}{\partial x^2}+\frac{\partial^2 \phi}{\partial y^2}=0=\frac{\partial^2 \psi}{\partial x^2}+\frac{\partial^2 \psi}{\partial y^2}$ or $\nabla^2 \phi = \nabla^2 \psi = 0$ where $\nabla^2 = \frac{\partial^2}{\partial x^2}+\frac{\partial^2}{\partial y^2}$	A flow described by a unique velocity potential is irrotational.
Equation of motion in 2-D	$\frac{\partial u}{\partial t}+u\frac{\partial u}{\partial x}+v\frac{\partial u}{\partial y}=\frac{1}{\rho}\left(X-\frac{\partial p}{\partial x}\right)$ $\frac{\partial v}{\partial t}+u\frac{\partial v}{\partial x}+v\frac{\partial v}{\partial t}=\frac{1}{\rho}\left(Y-\frac{\partial p}{\partial y}\right)$	The principle of force = mass × acceleration (Newton's law of motion) applies to fluids and fluid particles.
Equation of continuity in 2-D (incompressible flow)	$\frac{\partial u}{\partial x}+\frac{\partial v}{\partial y}=0$ or, in polar, $\frac{q_n}{r}+\frac{\partial q_n}{\partial r}+\frac{1}{r}\frac{\partial q_t}{\partial \theta}=0$	If fluid velocity increases in the *x* direction, it must decrease in the *y* direction.
Equation of vorticity	$\frac{\partial v}{\partial x}-\frac{\partial u}{\partial y}=\varsigma$ or, in polar, $\varsigma=\frac{q_t}{r}+\frac{\partial q_t}{\partial r}-\frac{1}{r}\frac{\partial q_n}{\partial \theta}$	A rotating or spinning element of fluid can be investigated by assuming it is a solid.
Stream function ψ (incompressible flow)	Velocity at a point is given by $u=\frac{\partial \psi}{\partial y} \quad v=-\frac{\partial \psi}{\partial x}$	ψ is the stream function. Lines of constant ψ give the flow pattern of a fluid stream.
Velocity potential ϕ (irrotational 2-D flow)	Velocity at a point is given by $u=\frac{\partial \phi}{\partial x} \quad v=\frac{\partial \phi}{\partial y}$	ϕ is defined as $\phi=\int_{op} q \cos \beta \, ds$

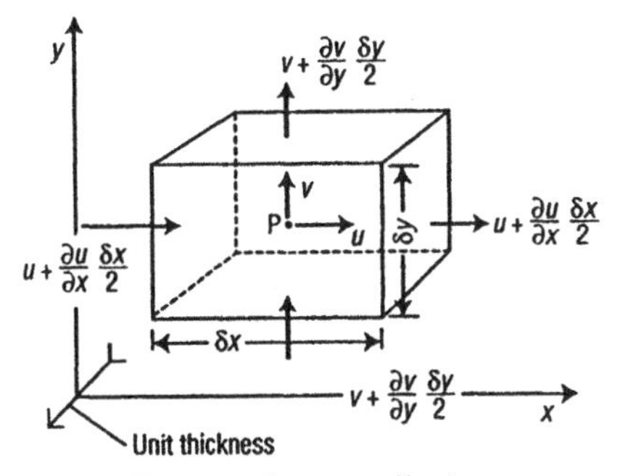

Rectangular co-ordinates

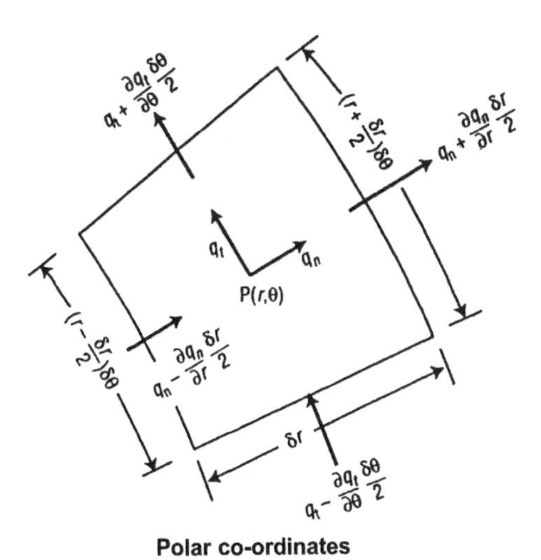

Polar co-ordinates

Fig. 6.2 Two-dimensional flow

The Navier–Stokes equations

The Navier–Stokes equations are written as

$$\rho\left(\frac{\partial u}{\partial t}+u\frac{\partial u}{\partial x}+v\frac{\partial u}{\partial y}\right)=\rho X-\frac{\partial p}{\partial x}+\mu\left(\frac{\partial^2 u}{\partial x^2}+\frac{\partial^2 u}{\partial y^2}\right)$$

$$\underbrace{\rho\left(\frac{\partial v}{\partial t}+u\frac{\partial v}{\partial x}+v\frac{\partial v}{\partial y}\right)}_{\text{Inertia term}}=\underbrace{\rho Y}_{\text{Body force term}}-\underbrace{\frac{\partial p}{\partial y}}_{\text{Pressure term}}+\underbrace{\mu\left(\frac{\partial^2 v}{\partial x^2}+\frac{\partial^2 v}{\partial y^2}\right)}_{\text{Viscous term}}$$

Sources and sinks

A 'source' is an arrangement where a volume of fluid, $+q$, flows out evenly from an origin toward the periphery of an (imaginary) circle around it. If q is negative, such a point is termed a 'sink' (see Fig. 6.3). If a source and sink of equal strength have their extremities infinitesimally close to each other, while increasing the strength, this is termed a 'doublet'.

6.3 Flow regimes

General descriptions

Flow regimes can be generally described as follows (see Fig. 6.4):

- Steady flow — Flow parameters at any point do not vary with time (even though they may differ between points).
- Unsteady flow — Flow parameters at any point vary with time.
- Laminar flow — Flow which is generally considered smooth, i.e. not broken up by eddies.
- Turbulent flow — Non-smooth flow in which any small disturbance is magnified, causing eddies and turbulence.
- Transition flow — The condition lying between laminar and turbulent flow regimes.

Reynolds number

Reynolds number is a dimensionless quantity that determines the nature of flow of fluid over a surface.

$$\text{Reynolds number } (Re)=\frac{\text{Inertia forces}}{\text{Viscous forces}}=\frac{\rho VD}{\mu}=\frac{VD}{\nu}$$

where

ρ = density
μ = dynamic viscosity
ν = kinematic viscosity
V = velocity
D = effective diameter

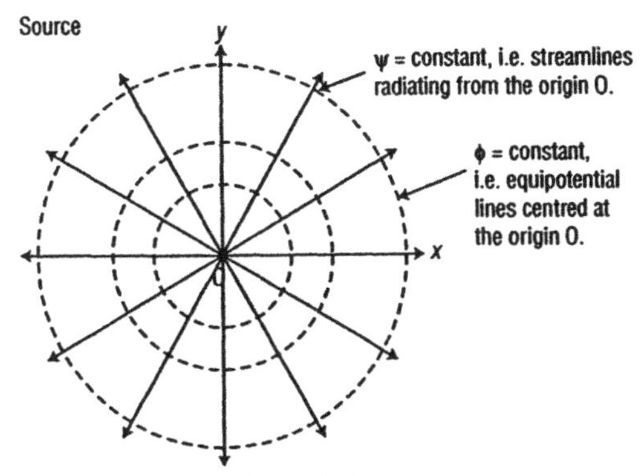

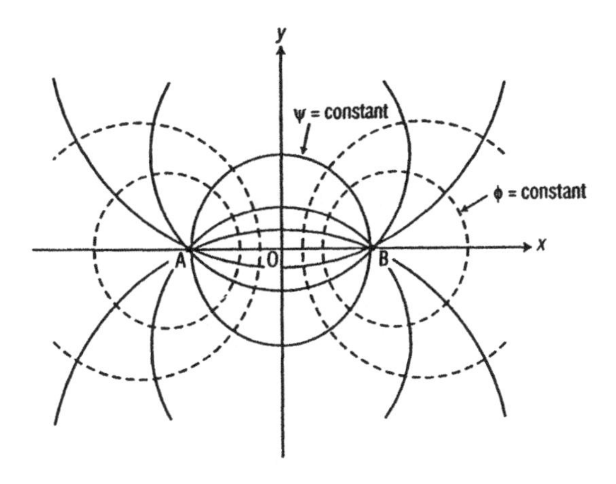

Fig. 6.3 Sources and sinks

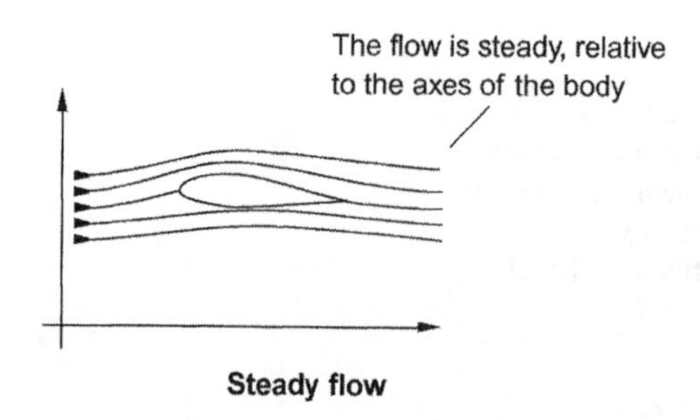

Steady flow

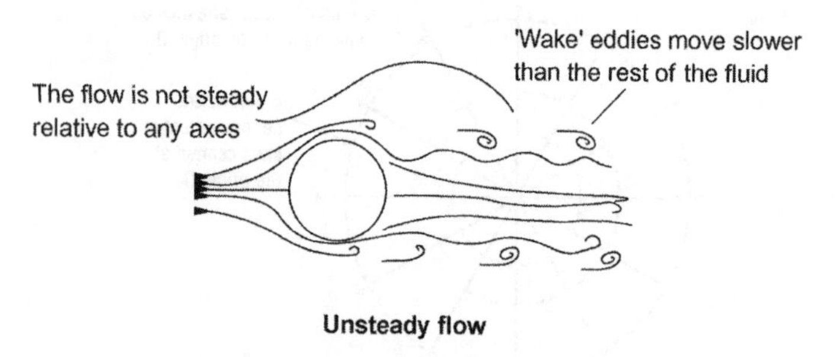

Unsteady flow

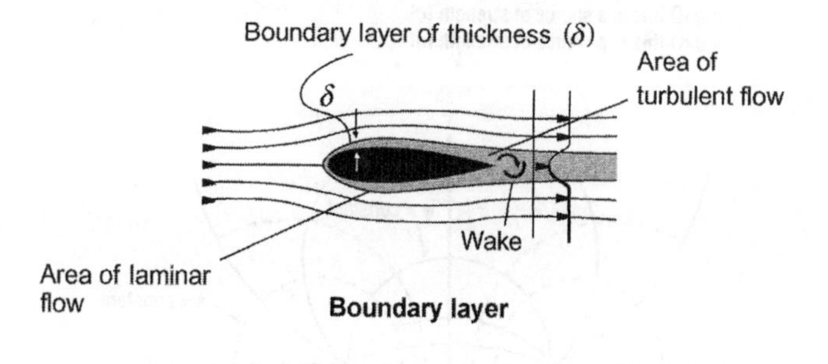

Boundary layer

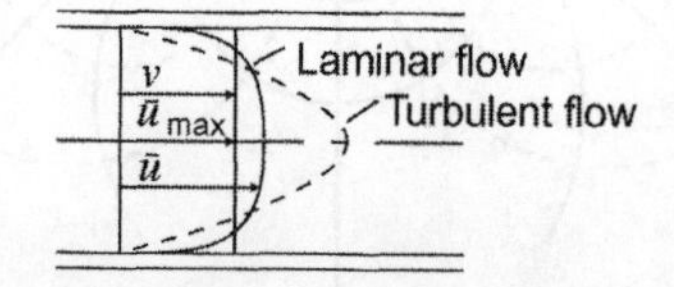

Velocity distributions in laminar and turbulent flow

Fig. 6.4 Flow regimes

Low Reynolds numbers (below about 2000) result in laminar flow.
High Reynolds numbers (above about 2300) result in turbulent flow.
Values of Re for $2000 < Re < 2300$ are generally considered to result in transition flow. Exact flow regimes are difficult to predict in this region.

6.4 Boundary layers

Figure 6.5 shows boundary layer velocity profiles for dimensional and non-dimensional cases. The non-dimensional case is used to allow comparison between boundary layer profiles of different thickness.

Definitions

- The *boundary layer* is the region near a surface or wall where the movement of a fluid flow is governed by frictional resistance.
- The *main flow* is the region outside the boundary layer that is not influenced by frictional resistance and can be assumed to be 'ideal' fluid flow.
- *Boundary layer thickness*. The thickness of the boundary layer is conventionally taken as the perpendicular distance from the surface of a component to a point in the flow where the fluid has a velocity equal to 99 per cent of the local mainstream velocity.

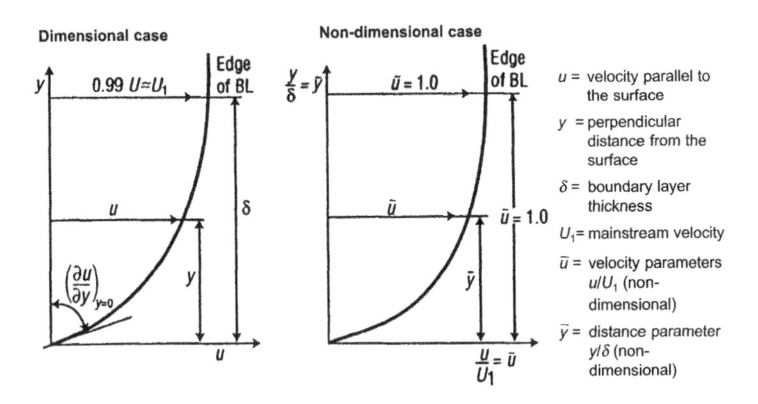

Fig. 6.5 Boundary layer velocity profiles

Some boundary layer equations

Boundary layer equations of turbulent flow

$$\rho\left(\bar{u}\frac{\partial \bar{u}}{\partial x}+\frac{\partial \bar{u}}{\partial y}\right)=-\frac{\partial \bar{p}}{\partial x}+\frac{\partial \tau}{\partial y}$$

$$\tau=\mu\frac{\partial \bar{u}}{\partial y}-\overline{\rho u'v'}$$

$$\frac{\partial \bar{p}}{\partial y}=0$$

$$\frac{\partial \bar{u}}{\partial x}+\frac{\partial \bar{v}}{\partial y}=0$$

6.5 Isentropic flow

For flow in a smooth pipe with no abrupt changes of section:

- Continuity equation

$$\frac{\mathrm{d}\rho}{\rho}+\frac{\mathrm{d}u}{u}+\frac{\mathrm{d}A}{A}=0$$

- Equation of momentum conservation

$$-\mathrm{d}p\,A=(A\rho u)\mathrm{d}u$$

- Isentropic relationship

$$p=c\rho^k$$

- Sonic velocity

$$a^2=\frac{\mathrm{d}p}{\mathrm{d}\rho}$$

These lead to an equation being derived on the basis of mass continuity, i.e.

$$\frac{\mathrm{d}\rho}{\rho}=-M^2\frac{\mathrm{d}u}{u}$$

or

$$M^2=-\frac{\mathrm{d}\rho}{\rho}\Big/\frac{\mathrm{d}u}{u}$$

Table 6.4 shows equations relating to convergent and convergent–divergent nozzle flow.

Table 6.4 Isentropic flows

Pipe flows	$\frac{-d\rho}{\rho} \Big/ \frac{du}{u} = M^2$
Convergent nozzle flows	Flow velocity $u = \sqrt{\left\{2\left(\frac{k}{k-1}\right)\left(\frac{p_o}{\rho_o}\right)\left[1 - \frac{p}{p_o}^{\frac{k-1}{k}}\right]\right\}}$ Flowrate $m = \rho u A$
Convergent–divergent nozzle flows	Area ratio $\frac{A}{A^*} = \frac{\left(\frac{2}{k+1}\right)^{\frac{1}{(k-1)}}\left(\frac{p_o}{p}\right)^{\frac{1}{k}}}{\sqrt{\left\{\frac{k+1}{k-1}\left[1 - \frac{p_o}{p}^{\frac{(1-k)}{k}}\right]\right\}}}$

6.6 Compressible one-dimensional flow

Basic equations for 1-D compressible flow are:
Euler's equation of motion in the steady state along a streamline

$$\frac{1}{\rho}\frac{dp}{ds} + \frac{d}{ds}\left(\frac{1}{2}u^2\right) = 0$$

or

$$\int \frac{dp}{\rho} + \frac{1}{2}u^2 = \text{constant}$$

so

$$\frac{k}{k-1}RT + \frac{1}{2}u^2 = \text{constant}$$

$$\frac{p_0}{p} = \left(\frac{T_0}{T}\right)^{k/(k-1)} = \left(1 + \frac{k-1}{2}M^2\right)^{k/(k-1)}$$

where T_o = total temperature

6.7 Normal shock waves

One-dimensional flow

A shock wave is a pressure front that travels at speed through a gas. Shock waves cause an increase in pressure, temperature, density and entropy and a decrease in normal velocity.

Equations of state and equations of conservation applied to a unit area of shock wave give (see Fig. 6.6)

State

$$p_1/\rho_1 T_1 = p_2/\rho_2 T_2$$

Mass flow

$$\dot{m} = \rho_1 u_1 = \rho_2 u_2$$

Momentum

$$p_1 + \rho_1 u_1^2 = p_2 + \rho_2 u_2^2$$

Energy

$$c_p T_1 + \frac{u_1^2}{2} = c_p T_2 + \frac{u_2^2}{2} = c_p$$

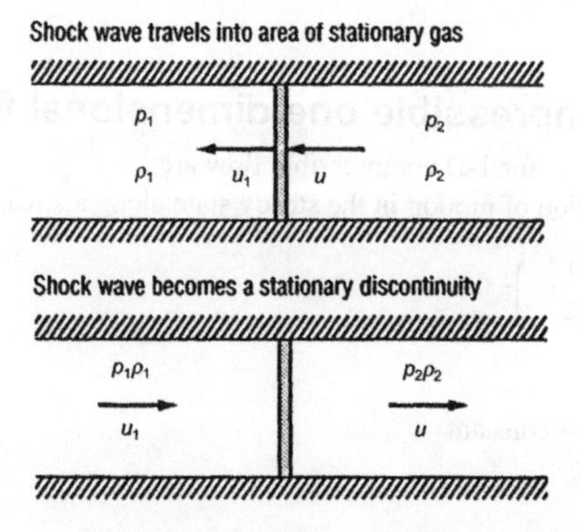

Fig. 6.6 Normal shock waves

Pressure and density relationships across the shock are given by the Rankine–Hugoniot equations

$$\frac{p_2}{p_1} = \frac{\dfrac{(\gamma+1)\rho_2}{(\gamma-1)\rho_1} - 1}{\dfrac{\gamma+1}{\gamma-1} - \dfrac{\rho_2}{\rho_1}}$$

$$\frac{\rho_2}{\rho_1} = \frac{\dfrac{(\gamma+1)p_2}{(\gamma-1)p_1} + 1}{\dfrac{\gamma+1}{\gamma-1} + \dfrac{p_2}{p_1}}$$

Static pressure ratio across the shock is given by

$$\frac{p_1}{p_2} = \frac{2\gamma\, M_2^2 - (\gamma-1)}{\gamma+1}$$

Temperature ratio across the shock is given by

$$\frac{T_2}{T_1} = \frac{p_2}{p_1} \Big/ \frac{\rho_2}{\rho_1}$$

$$\frac{T_2}{T_1} = \left(\frac{2\gamma\, M_1^2 - (\gamma-1)}{\gamma+1}\right)\left(\frac{2+(\gamma-1)M_1^2}{(\gamma+1)M_1^2}\right)$$

Velocity ratio across the shock is given from continuity by

$$u_2/u_1 = \rho_1/\rho_2$$

so

$$\frac{u_2}{u_1} = \frac{2+(\gamma-1)M_1^2}{(\gamma+1)M_1^2}$$

In axisymmetric flow the variables are independent of θ so the continuity equation can be expressed as

$$\frac{1}{R^2}\frac{\partial(R^2 q_R)}{\partial R} + \frac{1}{R\sin\varphi}\frac{\partial(\sin\varphi\, q_\varphi)}{\partial\varphi} = 0$$

Similarly in terms of stream function ψ

$$q_R = \frac{1}{R^2\sin\varphi}\frac{\partial\psi}{\partial\varphi}$$

$$q_\varphi = -\frac{1}{R\sin\varphi}\frac{\partial\psi}{\partial R}$$

The pitot tube equation

An important criterion is the Rayleigh supersonic pitot tube equation (see Fig. 6.7).

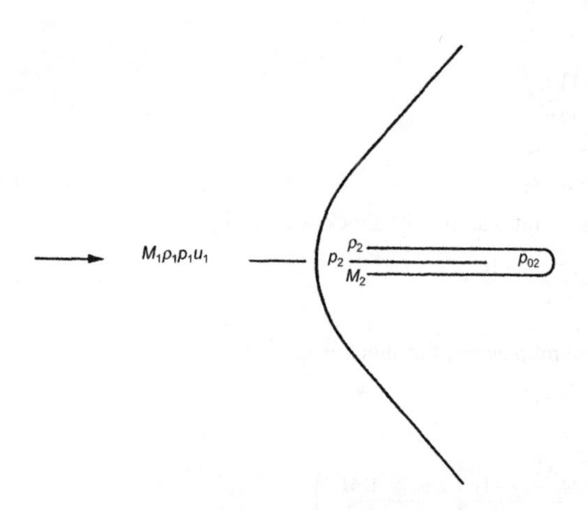

Fig. 6.7 Pitot tube

Pressure ratio

$$\frac{p_{02}}{p_1} = \frac{\left[\dfrac{\gamma+1}{2} M_1^2\right]^{\gamma/(\gamma-1)}}{\left[\dfrac{2\gamma M_1^2 - (\gamma-1)}{\gamma+1}\right]^{1/(\gamma-1)}}$$

6.8 Axisymmetric flows

Axisymmetric potential flows occur when bodies such as cones and spheres are aligned into a fluid flow. Figure 6.8 shows the layout of spherical co-ordinates used to analyse these types of flow.

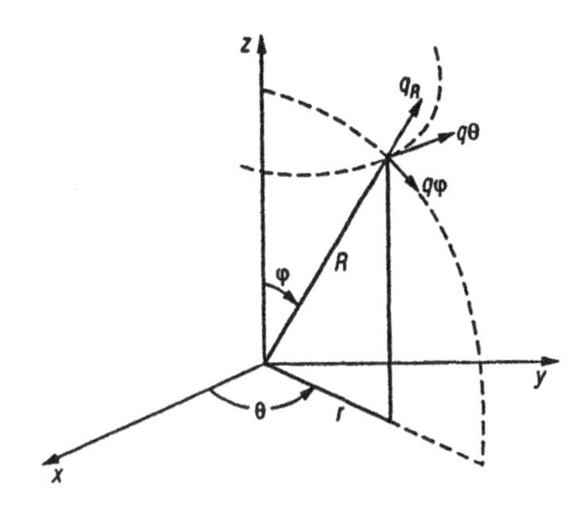

Fig. 6.8 Axisymmetric flows

Relationships between the velocity components and potential are given by

$$q_R = \frac{\partial \phi}{\partial R} \quad q_\theta = \frac{1}{R\sin\varphi}\frac{\partial \phi}{\partial \theta} \quad q_\varphi = \frac{1}{R}\frac{\partial \phi}{\partial \varphi}$$

6.9 Drag coefficients

Figure 6.9 shows drag types and 'rule of thumb' coefficient values.

Shape	*Dimensional ratio*	*Datum area, A*	*Approximate drag coefficient, C_D*
Cylinder (flow direction) 	$l/d = 1$		0.91
	2		0.85
	4	$\frac{\pi}{4}d^2$	0.87
	7		0.99
Cylinder (right angles to flow) 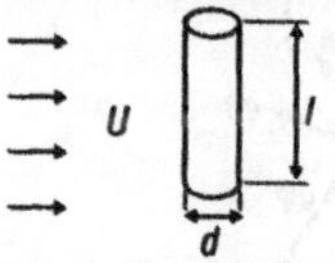	$l/d = 1$		0.63
	2		0.68
	5		0.74
	10	dl	0.82
	40		0.98
	∞		1.20
Hemisphere (bottomless) 	I	$\frac{\pi}{4}d^2$	0.34
	II		1.33
Cone 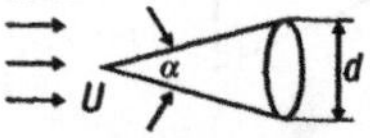	a = 60°	$\frac{\pi}{4}d^2$	0.51
	a = 30°		0.34
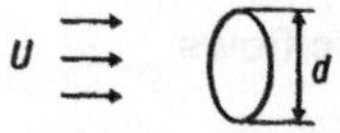		$\frac{\pi}{4}d^2$	1.2

Fig. 6.9 Drag coefficients

CHAPTER 7

Centrifugal Pumps

7.1 Symbols

Figure 7.1 shows some typical symbols used in schematic process and instrumentation diagrams (PIDs) incorporating items of rotating equipment (including pumps, fans, and compressors).

7.2 Centrifugal pump types

There are several hundred identifiable types of pump design tailored for varying volume throughputs and delivery heads, and including many specialized designs for specific fluid applications. The most common type, accounting for perhaps 80 per cent of fluid transfer applications, is the broad 'centrifugal pump' category.

There is a wide variety of centrifugal pump designs. Figure 7.2 shows some typical examples.

Figure 7.2(a) shows a back pull-out version of a basic, single-stage, centrifugal design. This allows the rotor to be removed towards the motor without disturbing the suction or delivery pipework. This type is commonly used for pumping of acids or hazardous fluids in the chemical and petrochemical industry.

Figure 7.2(b) shows a standard, horizontal, multi-stage, centrifugal design with a balance disc to enable axial thrust to be hydraulically balanced. The most common application is for high-pressure boiler feed water.

Figure 7.2(c) shows a horizontal, multi-stage, centrifugal pump using the side-channel principle combined with a radial flow suction stage impeller. This special impeller, with an axial inlet branch, is arranged upstream of the open star vane impellers. In this way, combination pumps are obtained

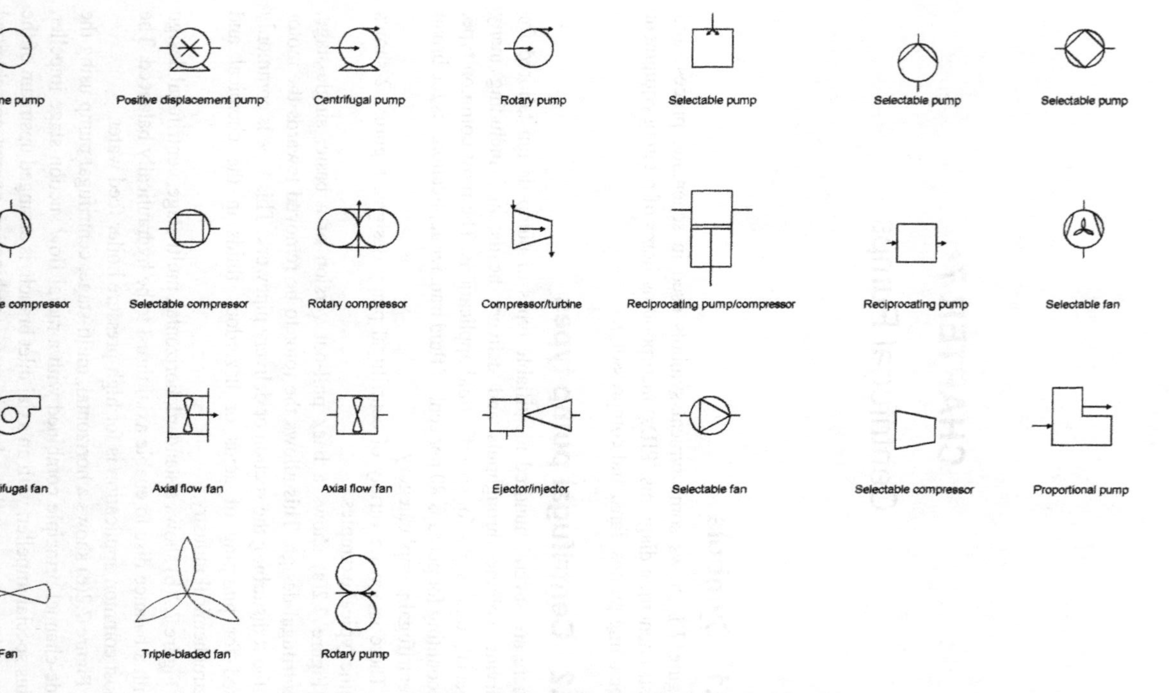

Fig. 7.1 Some typical PID symbols (Courtesy MS Visio)

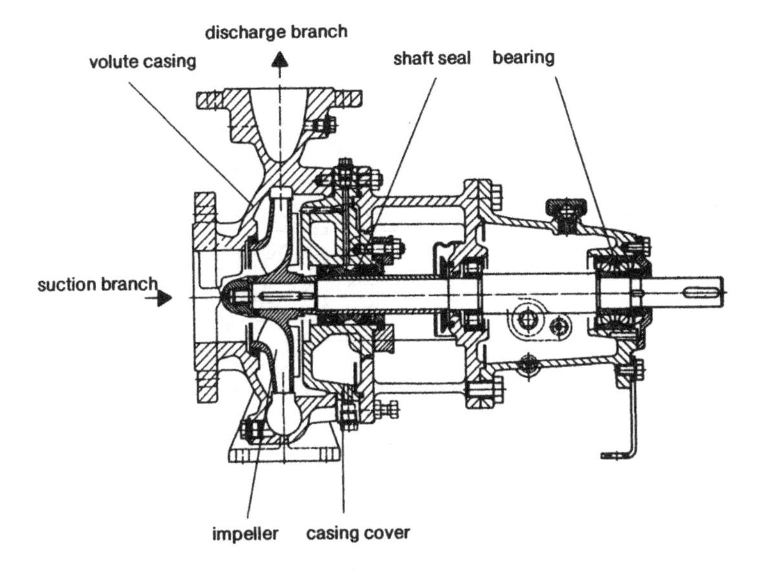

Fig. 7.2(a) Single-stage, centrifugal pump (back pull-out design)

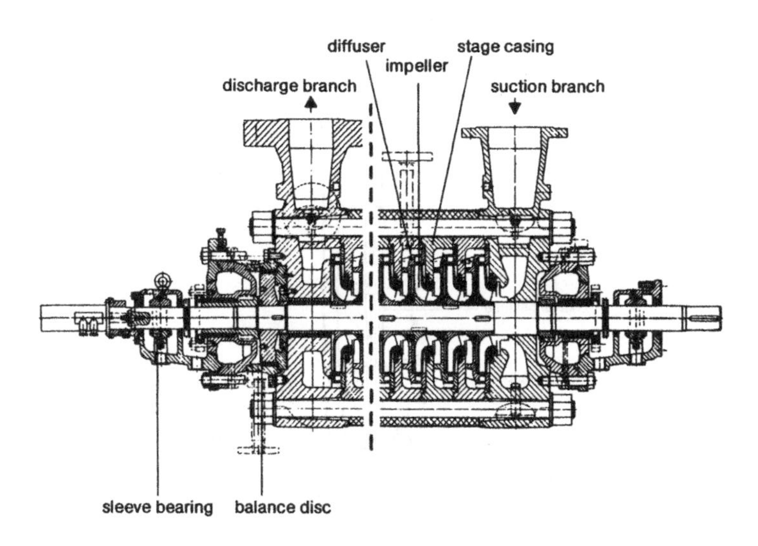

Fig. 7.2(b) Horizontal, multi-stage, centrifugal pump

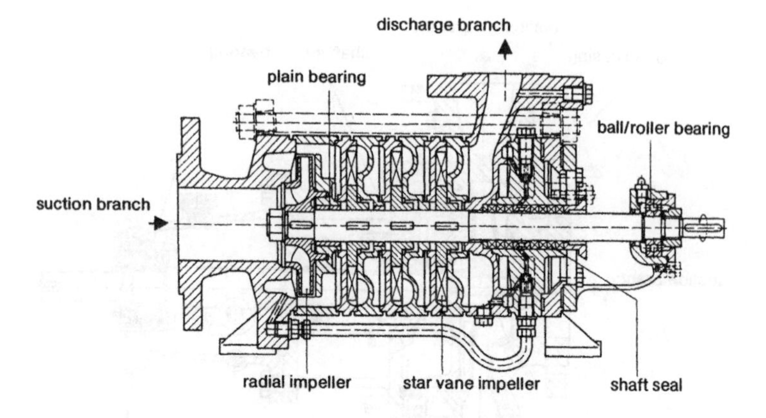

Fig. 7.2(c) Side-channel, multi-stage, centrifugal pump (self-priming with low NPSH inlet stage)

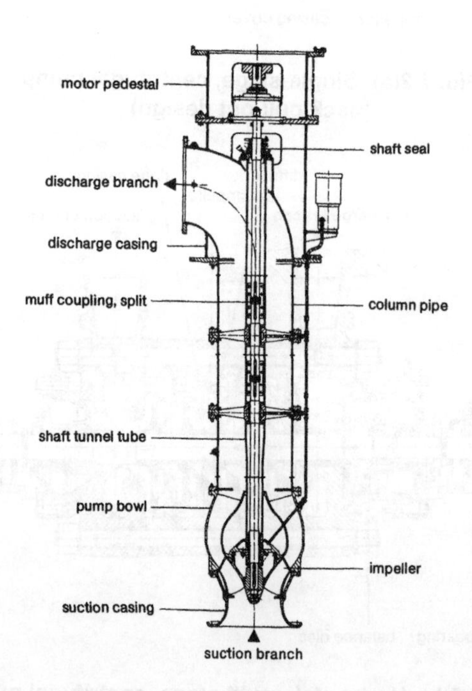

Fig. 7.2(d) Vertical, centrifugal, mixed-flow pump

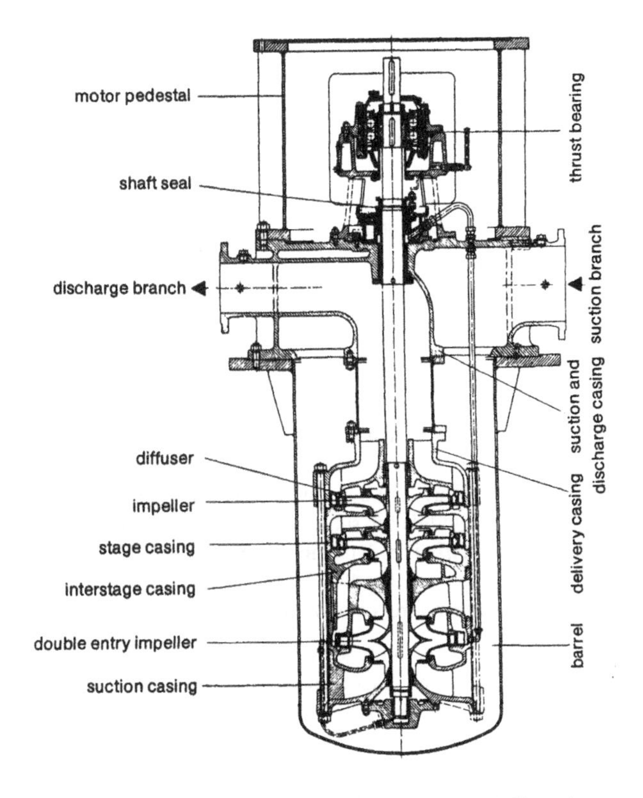

Fig. 7.2(e) Vertical barrel, multi-stage, centrifugal pump

which, in addition to the specific features of the side-channel principle, e.g. self-priming, gas handling capacity, and high head per stage, have a very low Net Positive Suction Head (NPSH) requirement. Applications include boiling liquids with low suction heads, condensate, boiler feed water, liquid gas, and refrigerants.

Figures 7.2(d) and (e) show vertical, multi-stage designs.

7.3 Pump performance

There are many pump performance parameters, some of which are complex and may be presented in a non-dimensional format.

Volume flowrate q

Flowrate is the first parameter specified by the process designer who bases the pump requirement on the flowrate that the process needs to function. This 'rated' flowrate is normally expressed in volume terms and is represented by the symbol q, with units of m^3/s.

Head H

Once rated flowrate has been determined, the designer then specifies a total head H required at this flowrate. This is expressed in metres and represents the usable mechanical work transmitted to the fluid by the pump.

In general, the usable mechanical energy of a liquid is the sum of energy of position, pressure energy, and dynamic energy. The pressure energy per unit of weight of the liquid that is subject to the static pressure p is termed the 'pressure head' $p/(\rho g)$. The dynamic energy of the liquid, per unit of weight, is termed the 'velocity head' $v^2/2g$.

The total head H is, therefore, composed of

$z_d - z_s$ = difference of altitude (i.e. of height) between the outlet branch and the inlet branch of a pump

$\frac{p_d - p_s}{\rho g}$ = difference of pressure head of the liquid between the outlet branch and the inlet branch of a pump

$\frac{v_d^2 - v_s^2}{2g}$ = difference of velocity head of the liquid between the outlet branch and the inlet branch of the pump

From the above, the total head of the pump is

$$H = (z_d - z_s) + \frac{p_d - p_s}{\rho g} + \frac{v_d^2 - v_s^2}{2g}$$

Together q and H define the 'duty point', the core FFP criterion.

Net Positive Suction Head (NPSH)

NPSH is slightly more difficult to understand. Essentially, it is a measure of the pump's ability to avoid cavitation in its inlet (suction) region. This is done by maintaining a pressure excess above the relevant vapour pressure in this inlet region. This pressure excess keeps the pressure above that at which cavitation will occur. Acceptance guarantees specify a maximum NPSH required. The unit is metres.

The reference plane for the NPSH value is defined by the horizontal plane that passes through the centre of the circle. This is determined by the most extreme points of the leading edge of the blades (Fig. 7.3). In the case of double-entry pumps with a shaft that is not horizontal, the impeller inlet located at the higher level is the determining factor. For pumps with a horizontal shaft the reference plane lies in the centre of the shaft. For pumps with a vertical shaft or a shaft that is inclined to the vertical, the position of the impeller inlet, and hence the reference plane for the NPSH value, cannot be determined from the outside and it has to be given by the manufacturer.

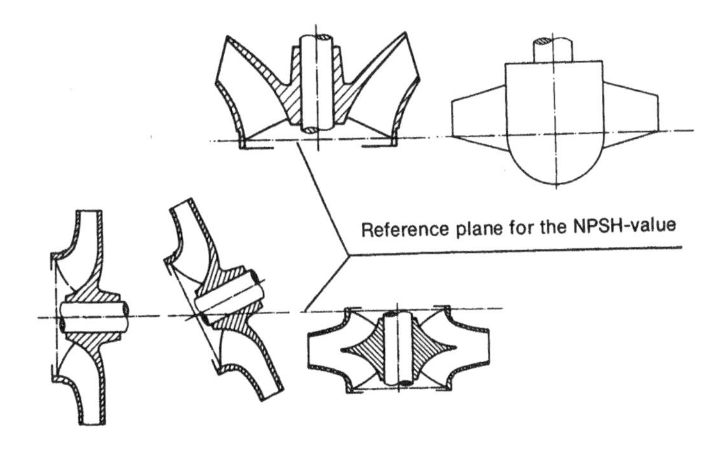

Fig. 7.3 Reference plane for NPSH

Available NPSH of an installation

The available NPSH value ($NPSH_{avail}$) of an installation (see Fig. 7.4) is the difference between the total head (static pressure head $(p_s + p_b)/\rho g$ plus velocity head $v_s^2/2g$) and the vapour pressure head $p_D/\rho g$ referred to the reference plane for the NPSH value

$$NPSH_{avail} = \frac{p_d + p_b}{\rho g} + \frac{v_s^2}{2g} - \frac{p_D}{\rho g} + z_s'$$

z_s' = difference of level between the centre of the inlet branch of the pump and the reference plane for the NPSH value.

z_s' is +ve if the reference plane for the NPSH value lies below the centre of the pump inlet.

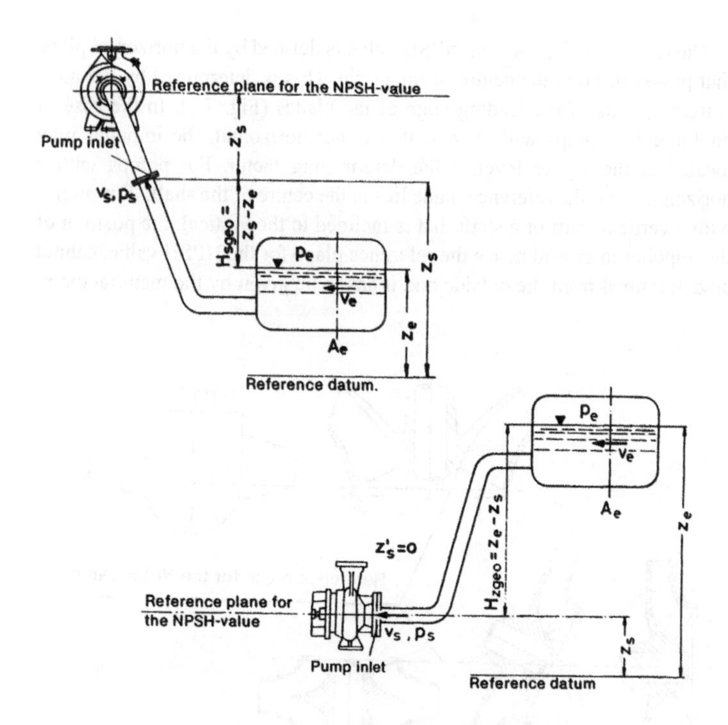

Fig. 7.4 Available NPSH of a system

z_s' is –ve if the reference plane for the NPSH value lies above the centre of the pump inlet (see Fig. 7.3).

z_s' is zero if the reference plane for the NPSH value lies at the same level as the centre of the pump inlet.

The total head at the centre of the inlet connection of the pump can be derived from the total head at the system inlet.

If a static suction lift $H_{sgeo} = z_s - z_e$ has to be taken into account, then

$$\frac{p_s + p_b}{\rho g} + \frac{v_s^2}{2g} = \frac{p_e + p_b}{\rho g} + \frac{v_e^2}{2g} - H_{sgeo} - H_{vs}$$

and hence the available NPSH value of the installation is

$$\text{NPSH}_{avail} = \frac{p_e + p_b - p_D}{\rho g} + \frac{v_e^2}{2g} - H_{sgeo} - H_{vs} + z_s'$$

or where the pump draws from an open container ($p_e = 0$), i.e. under suction lift conditions

$$\mathrm{NPSH}_{\mathrm{avail}} = \frac{p_b - p_D}{\rho g} + \frac{v_e^2}{2g} - H_{\mathrm{sgeo}} - H_{\mathrm{vs}} + z_s{}'$$

If a static suction head $H_{zgeo} = -H_{sgeo} = z_e - z_s$ is given, the available NPSH value of the installation is

$$\mathrm{NPSH}_{\mathrm{avail}} = \frac{p_e + p_b - p_D}{\rho g} + \frac{v_e^2}{2g} + H_{\mathrm{sgeo}} - H_{\mathrm{vs}} + z_s{}'$$

or where the pump delivers from an open container ($p_e = 0$), i.e. under suction head conditions

$$\mathrm{NPSH}_{\mathrm{avail}} = \frac{p_b - p_D}{\rho g} + \frac{v_e^2}{2g} + H_{\mathrm{sgeo}} - H_{\mathrm{vs}} + z_s{}'$$

In practice the velocity head $v_e^2/2g$ in the container on the suction side of the pump is small enough to be neglected.

For trouble-free operation of a pump the condition $\mathrm{NPSH}_{avail} \geq \mathrm{NPSH}_{req}$ has to be satisfied. For reasons of safety and to cover transient conditions, it is recommended that an excess of approximately 0.5 m is provided, i.e.

$$\mathrm{NPSH}_{avail} \geq \mathrm{NPSH}_{req} + \text{approx. } 0.5 \text{ m}$$

Other criteria

- Pump efficiency (η per cent): the efficiency with which the pump transfers mechanical work to the fluid.
- Power (P) Watts consumed by the pump.
- Noise and vibration characteristics.

It is normal practice for the above criteria to be expressed in the form of 'acceptance guarantees' for the pump.

7.4 Pump characteristics

Figure 7.5 shows a typical centrifugal pump characteristic.

Fluid Equipment

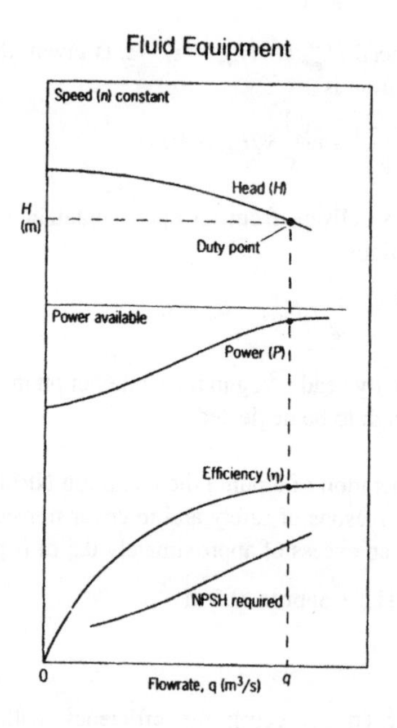

Fig. 7.5 Typical centrifugal pump characteristics

The q/H curve

The test is carried out at a nominally constant speed, and the head H decreases as flowrate q increases, giving a negative slope to the curve. Note how the required 'duty point' is represented, and how the required pump power and efficiency change as flowrate varies.

The NPSH (required) curve

NPSH needs two different sets of axes to describe it fully. The lower curve in Fig. 7.5 shows how NPSH 'required' to maintain full head performance rises with increasing flowrate. Note, however, that this curve is not obtained directly from the q/H test; it is made up of three or four points, each point being obtained from a separate NPSH test at a different constant q. This is normally carried out after the q/H test. In the NPSH test, the objective is for the pump to maintain full head performance at an NPSH equal to or less than a maximum 'guarantee' value.

Table 7.1 shows indicative values for a large circulating water pump.

Table 7.1 Typical acceptance guarantee schedule

Rated speed n	740 r/min	
Rated flowrate q	0.9 m^3/s	together, these define the 'duty point'
Rated total head H	60 m	
Rated efficiency	80 per cent at duty point	
Absorbed power	660 kW at duty point	
NPSH	Maximum 6 m at impeller eye for 3 per cent total head drop	
Vibration	Vibration measured at the pump bearing shall not exceed 2.8 mm/s r.m.s. at the duty point	
Noise	Maximum allowable level = 90 dB(A) at duty point (at agreed measuring locations)	

Now the specification states:

- *Tolerances*: ± 1.5 per cent on head H and ± 2 per cent on flow q (these are typical, but can be higher or lower, depending on what the designer wants) *but* + 0 on NPSH.
- *The acceptance test standard*: e.g. ISO 3555. This is important; it tells you a lot about test conditions and which measurement tolerances to take into account when you interpret the curves.

7.5 Specifications and standards

Some well-proven centrifugal pump test standards are:

- ISO 2548 (identical to BS 5316 Part 1) is for 'Class C' levels of accuracy. This is the least accurate class and has the largest allowable 'measurement tolerances' which are applied when drawing the test curves, and hence the largest 'acceptance' tolerances on q and H.
- ISO 3555 (identical to BS 5316 Part 2) is for 'Class B' levels of accuracy, with tighter test tolerances than for Class C.
- ISO 5198 (identical to BS 5316 Part 3) is for 'Class A' (or 'precision') levels of accuracy. This is the most stringent test with the tightest tolerances.

- DIN 1944 *Acceptance tests for centrifugal pumps*. This is structured similarly to BS 5316 and has three accuracy classes, in this case denoted Class I, II, or III.
- API 610 *Centrifugal pumps for general refinery service*. This is a more general design-based standard.
- ISO 1940/1 (identical to BS 6861 Part 1) is commonly used to define dynamic balance levels for pump impellers.
- VDI 2056 is commonly used to define bearing housing or pump casing vibration. A more complex method, measuring shaft vibration, is covered by ISO 7919-1 (similar to BS 6749 Part 1).
- DIN 1952 and VDI 2040 are currently withdrawn standards but are still in common use to specify methods of flowrate q measurement.

7.6 Test procedures and techniques

Figure 7.6 shows a basic centrifugal pump test circuit. The test is carried out as follows:

Step 1. *The q/H test*

The first set of measurements is taken at duty point (100 per cent q). The valve is opened to give a flowrate greater than the duty flow (normally 120 or 130 per cent q) and further readings taken. The valve is then closed in a series of steps, progressively decreasing the flow (note that we are moving from right to left on the q/h characteristic). With some pumps, the final reading can be taken with the valve closed, i.e. the $q = 0$, 'shut-off condition'. The procedure is now: (Fig. 7.7)

- draw in the test points on the q/H axes;
- using the measurement accuracy levels given for the class of pump, draw in the q/H measured band;
- add the rectangle, which describes the tolerances allowed by the acceptance guarantee on total head H and flowrate q; ISO 3555 indicates tolerances of $\pm$ 2 per cent H and $\pm$ 4 per cent q;
- if the q/H band intersects or touches the rectangle then the guarantee has been met; note that the rectangle does not have to lie fully within the q/H band to be acceptable.

It is not uncommon to find different interpretations placed on the way in which ISO 3555 specifies 'acceptance' tolerances. The standard clearly specifies measurement accuracy levels $\pm$ 2 per cent q, $\pm$ 1.5 per cent H, but later incorporates these into a rigorous method of verifying whether the test curve meets the guarantee by using the formula for an ellipse (effectively allowing an elliptical tolerance 'envelope' around each measured point), specifying values of 2 per cent H and 4 per cent q to be used as the major axes lengths of the ellipse.

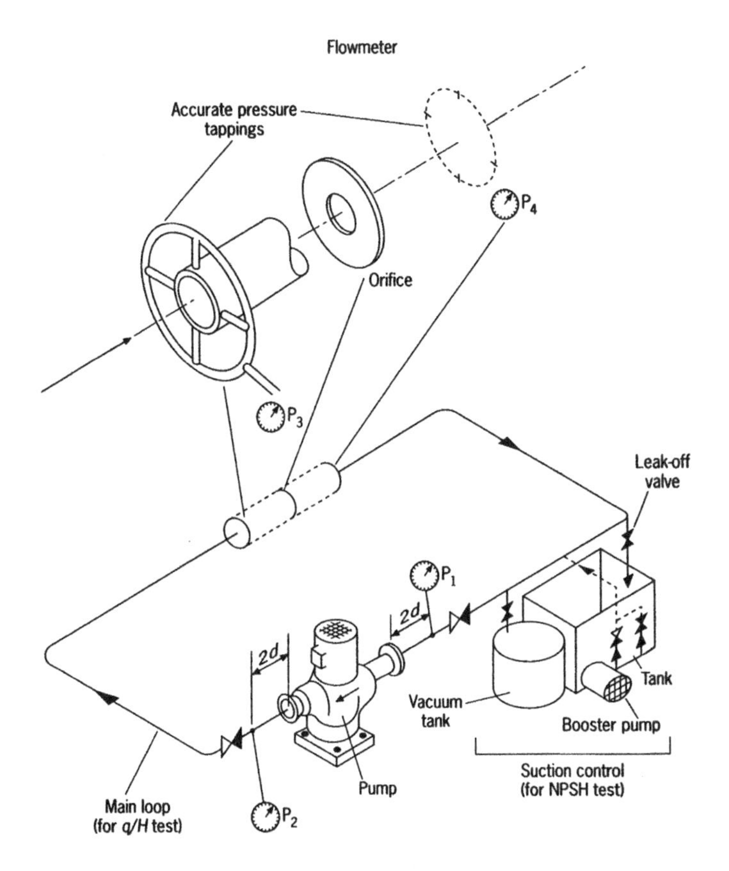

Fig. 7.6 Centrifugal pump test circuit

Step 2. *The efficiency test*
The efficiency guarantee is checked using the same set of test measurements as the q/H test. Pump efficiency is shown plotted against q as in Fig. 7.5. In most cases, the efficiency guarantee will be specified at the rated flowrate (q).

Step 3. *Noise and vibration measurements*
Vibration levels for pumps are normally specified at the duty (100 per cent q) point. The most common method of assessment is to measure the vibration level at the bearing housings using the methodology proposed by

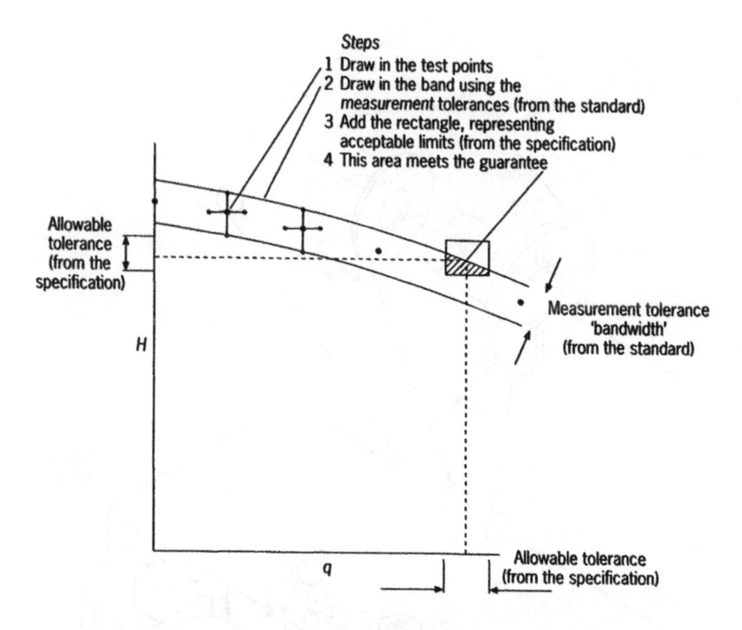

Fig. 7.7 Compliance with the *q/H* guarantee

VDI 2056. This approximates vibration at multiple frequencies to a single velocity (r.m.s.) reading. It is common for pumps to be specified to comply with VDI 2056 group T vibration levels, so a level of up to around 2.8 mm/s is acceptable. Some manufacturers scan individual vibration frequencies, normally multiples of the rotational frequency, to gain a better picture of vibration performance.

Pump noise is also measured at the duty point. It is commonly specified as an 'A-weighted sound pressure level' measured in dB(A) at the standard distance of 1 m from the pump surface.

Step 4. *The NPSH test*

These are two common ways of doing the NPSH test.

1. One can simply check that the pump performance is not impaired by cavitation at the specified *q/H* duty with the 'installed' NPSH of the test rig. This is a simple go/no-go test, applicable only for values of specified NPSH that can be built in to the test rig. It does not give an indication of any NPSH margin that exists, hence is of limited accuracy.

2. A comprehensive test technique is to explore NPSH performance more fully by varying the NPSH over a range and watching the effects. The most common method is the '3 per cent head drop' method shown in Fig. 7.8.

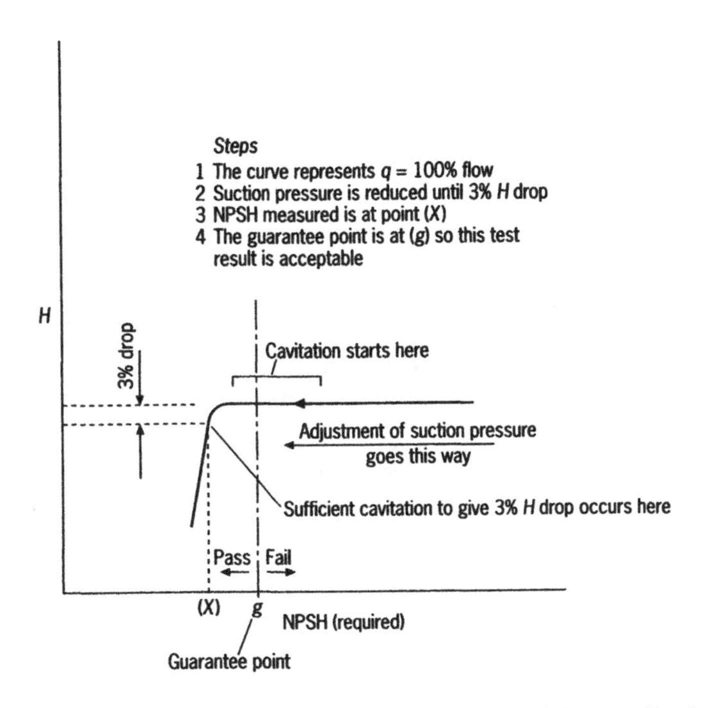
Steps
1 The curve represents q = 100% flow
2 Suction pressure is reduced until 3% H drop
3 NPSH measured is at point (X)
4 The guarantee point is at (g) so this test result is acceptable
H
3% drop
Cavitation starts here
Adjustment of suction pressure goes this way
Sufficient cavitation to give 3% H drop occurs here
Pass
Fail
(X)
g
NPSH (required)
Guarantee point

Fig. 7.8 Measuring NPSH – the 3 per cent head drop method

The test rig suction pressure control circuit is switched in, see Fig. 7.6, and the suction pressure reduced in a series of steps. For each step, the pump outlet valve is adjusted to keep the flowrate q at a constant value. The final reading is taken at the point where the pump head has decayed by at least 3 per cent. This shows that a detrimental level of cavitation is occurring and defines the attained NPSH value, as shown in Fig. 7.8. In order to be acceptable, this reading must be less than, or equal to, the maximum guarantee value specified. Strictly, unless specified otherwise, there is no 'acceptance' tolerance on NPSH, although note that ISO 3555 gives a *measurement* tolerance of ± 3 per cent or 0.15 m NPSH.

Corrections

Correction factors (applied to q, H, P, and NPSH) need to be used if the test speed of the pump does do not match the rated speed. They are:

- flow q (corrected) = q (measured) × (nsp/n)
- head H (corrected) = H (measured) × $(nsp/n)^2$
- power P (corrected) = P (measured) × $(nsp/n)^3$
- NPSH (corrected) = NPSH (measured) × $(nsp/n)^2$

  n = speed during the test

  nsp = rated speed

Table 7.2 shows some common practical problems and solutions that arise from centrifugal pump tests.

Table 7.2 Common problems in pump tests

Problems	*Corrective action*
The *q/H* characteristic is above and to the right of the guarantee point (i.e. too high).	For radial and mixed-flow designs, this is rectified by trimming the impeller(s). The *q/H* curve is moved down and to the left.
The *q/H* characteristic is 'too low' – the pump does not fulfil its guarantee requirement for *q* or *H*.	Often, up to 5% head increase can be achieved by fitting a larger diameter impeller. If this does not rectify the situation there is a hydraulic design fault, probably requiring a revised impeller design. Interim solutions can sometimes be achieved by: • installing flow-control or pre-rotation devices; • installing upstream throttles.
NPSH is well above the acceptance guarantee requirements.	This is most likely a design problem; the only real solution is to redesign.

Table 7.2 Cont.

Excessive vibration over the speed range.	The pump must be disassembled. First check the impeller dynamic balance using ISO 2373/BS 4999 part 142/IEC.42 or ISO 1940 for guidance. Check all the pump components for 'marring' and burrs: these are prime causes of inaccurate assembly. During re-assembly, check concentricities by measuring Total Indicated Runout (TIR) with a dial gauge.
Excessive vibration at rated speed.	Check the manufacturer's critical speed calculations. The first critical speed should be a minimum 15–20% *above the rated speed.* High vibration levels at discrete, rotational frequency is a cause for concern. A random vibration signature is more likely to be due to the effects of fluid turbulence.
Noise levels above the acceptance guarantee levels.	Pump noise is difficult to measure because it is masked by fluid flow noise from the test rig. If high noise levels are accompanied by vibration a stripdown and retest is necessary.

7.7 Pump specific speed n_s

Specific speed is a dimensionless characteristic relating to the shape of a pump impeller. In formal terms, it relates to the rotational speed of an impeller which provides a total head of 1 m at a volumetric flow rate of 1 m^3/s. From dynamic similarity, it can be shown that

$$n_s = n \cdot \frac{(q / q_s)^{1/2}}{(H / H_s)^{3/4}}$$

where

n is in r/min
q is in m^3/s
H is in m

Hence substituting $q_s = 1$ and $H_s = 1$ gives

$$n_s = n \cdot \frac{q^{1/2}}{H^{3/4}}$$

where q and H refer to the point of optimum efficiency of the impeller.

This formula can be expressed as a characteristic type number so that it remains non-dimensional, whatever system of units is used. Figure 7.9 shows some approximate design ranges for pump types based on their specific speed. Figure 7.10 shows the influences of specific speed on the shape of pump characteristic curves.

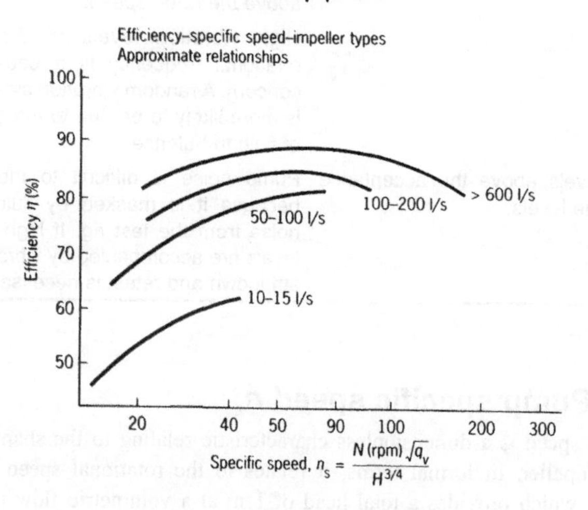

Fig. 7.9 Efficiency-specific speed-impeller types – approximate relationships

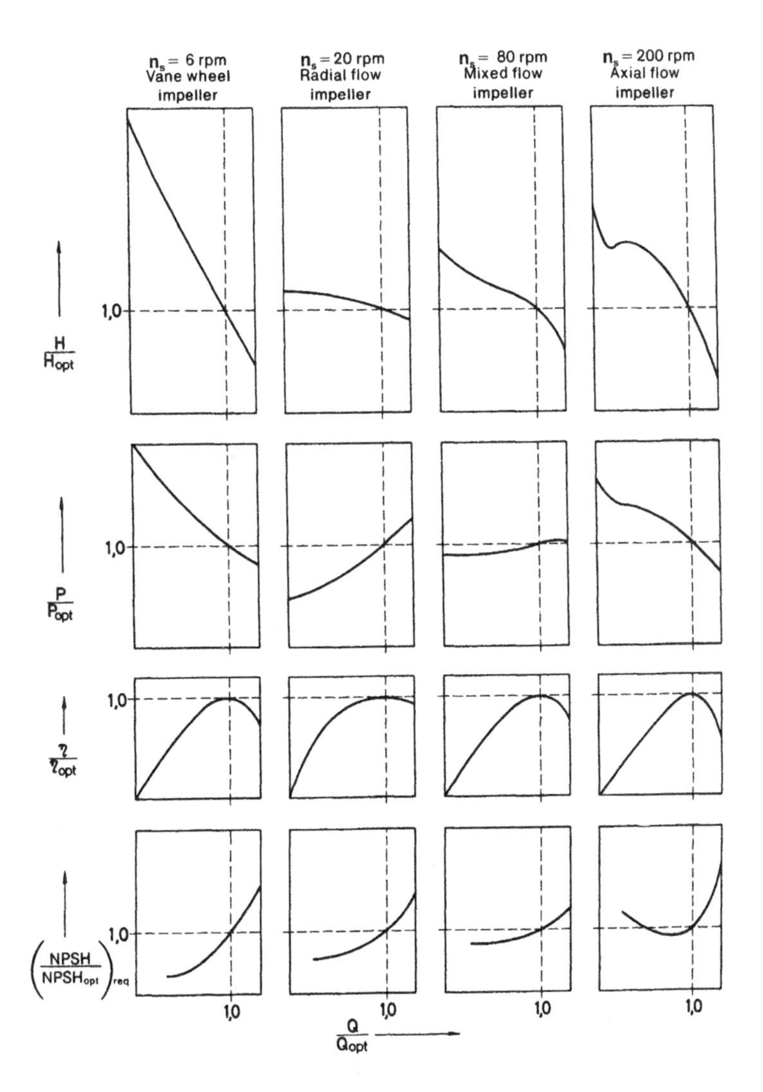

Fig. 7.10 The influence of specific speed on pump characteristic shape

7.8 Pump balancing

Technically, the balancing of pumps follows the same principles and standards used for other rotating equipment. However, there are a few points which are specific to pumps.

The rigid rotor assumption

Except in very unusual design circumstances, pumps are assumed to have rotors which behave as a rigid body, hence the standard ISO 1940 can be used. This applies both to pump designs in which the impeller is mounted between bearings and to those in which the impeller is overhung from a single bearing.

Static versus dynamic balancing

Rotating equipment can be balanced using either the 'static' method, which uses a single balancing correction plane, or by 'dynamic' balancing using two correction planes and taking into account the resulting couple imbalance that occurs. Pumps divide neatly into two categories based on the approximate ratio of the dimensions of the rotating parts.

- *Narrow 'high ratio' impellers.* If the diameter: width ratio of a pump's impeller is greater than six, then it is normal to use a simple static balancing technique, so balance correction is only carried out in one plane. There is a practical as well as a theoretical angle to this; it can be difficult to remove sufficient metal from a high ratio impeller without weakening the impeller itself.
- *Low ratio impellers.* When the impeller ratio is six or less, then two-plane dynamic balancing is used. This involves two separate allowable unbalance limits: one for static unbalance and one for dynamic unbalance, which takes account of the couple.

Impeller only versus assembly balance

Most pumps' impellers are balanced alone on the balancing machine, but there are occasions where the impeller and its rotating shaft are balanced in their assembled state. The rationale behind this approach is that the specified balance grade cannot be reached for the separate rotating components – but it is possible when they are assembled together.

Balance quality grade

European practice is to use ISO 1940 as the basis of specifying pump balance levels. Unfortunately the standard is not absolutely definitive in specifying acceptable levels of unbalance for pumps with various rotational speeds. Table 1 of the standard indicates an acceptance grade of G6.3 for pump impellers, but you can consider this as general guidance only – it is not necessarily applicable to all types of pump. Several major pump manufacturers use their own acceptance criteria based on a broad fitness-for-purpose assessment, but these may be changed to fit in with purchasers' specific requirements. These grades depend on rated speed – slow-speed pumps such as those for seawater circulation duties run at low speed and so can tolerate greater unbalance without suffering excessive stresses and resultant mechanical damage.

7.9 Balance calculations

The essence of balance calculations is incorporated in Table 1 of ISO 1940. This can be tricky to follow and it is difficult to obtain accurate readings from the logarithmic scales. A simpler method using calculation is shown in Fig. 7.11. The two main criteria are the allowable static residual unbalance U_s and the allowable couple unbalance U_c, which is itself a function of U_s and the physical dimensions of the pump. Figure 7.12 shows a typical calculation for a supported-bearing rotor balanced alone, i.e. without its shaft assembly.

The accuracy of balancing results, while important, is rarely a major issue when witnessing pump balancing tests in the works. The difference in allowable unbalance between grade G6.3 and G40 for instance is a factor of six or more, so small errors in balancing accuracy up to about ± 5 per cent can be treated as second order.

Test speed

There are no hard and fast rules about the test speed to be used during the balancing test. Theoretically, any speed up to the rated speed could be used, and give comparable unbalance readings, but practically the minimum acceptable test speed is governed by the sensitivity of the balancing machine. If the test speed is too low the machine will not give accurate results. The type of mandrel used to mount the impeller also has an effect on the test speed. Mandrels can be of either the fixed collar type, which fits accurately the impeller bore, or the 'expanding' type, which provides a universal fitting. Expanding mandrels cannot provide such good concentric accuracy and so can induce errors into the readings.

Mandrel accuracy

Whether a fixed collar or expanding mandrel is used, it must itself be balanced so that it does not induce errors into the readings of impeller

Static

Allowable static unbalance (U_S) is:

$$U_S = \frac{30.m.G}{\pi.N} \text{ (g.mm)}$$

This is equivalent to the $U_s = e_{per}M$ value in Table 1 of ISO 1940

Dynamic

Allowable dynamic (couple) unbalance is:

$$U_C = \frac{U_S.L}{2T} \text{ (g.mm)}$$

L = Bearing centre distance (mm)
T = Width across impeller shroud (mm)
m = Impeller mass (grammes)
M = Impeller mass (kg)
G = Balance grade (from ISO 1940)
N = Rated speed (r/min)
e_{per} = Residual unbalance (g.mm/kg)

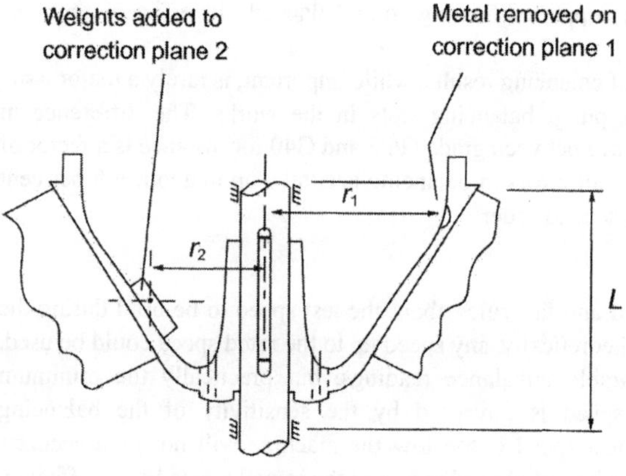

Weights added to correction plane 2

Metal removed on correction plane 1

r_1

r_2

L

Fig. 7.11 Pump impeller balancing

The balancing machine readout is:

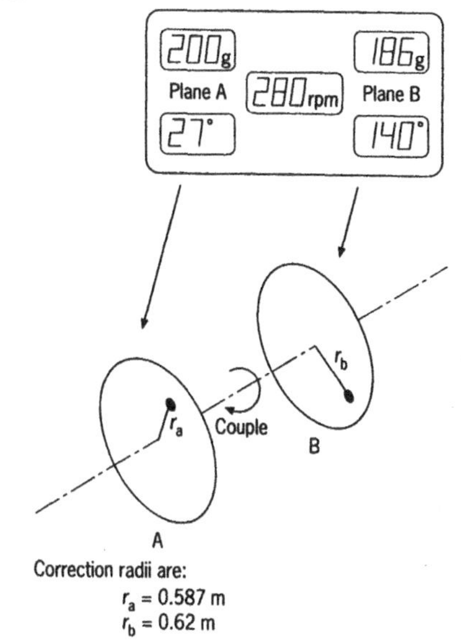

Correction radii are:

$r_a = 0.587$ m
$r_b = 0.62$ m

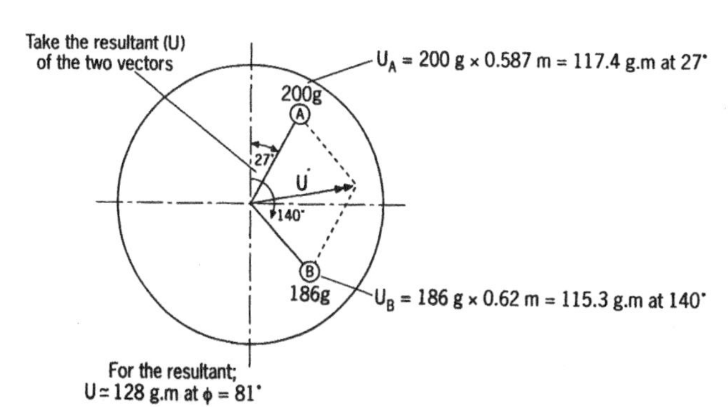

Show the results like this

Angle of resultant	U_{Max}	U_{Actual}
$\phi < 90°$	165g.m	128g.m
$\phi > 90°$	123g.m	–

This rotor is within limits

Fig. 7.12 Pump rotor balancing

unbalance. The acceptable balance grade for mandrels is normally taken as ISO 1940 G2.5, but in a practical works situation it is acceptable to use an approximate method, by measuring concentricities. This gives a first-order approximation of the mandrel balance grade – good enough for most purposes. Note how the typical acceptable concentricity limit (TIR) decreases as the pump rated speed increases.

Pump rated speed (r/min)	***Mandrel accuracy (TIR)***
Up to 1500	45 μm
1500–6000	20 μm
6000–7500	10 μm
above 7500	Mandrel not used

Note that expanding mandrels are not used for impellers with rated speeds above 1500 r/min.

Achieving impeller balance grade

Once the initial unbalance readings have been taken, the impeller (or the impeller/shaft assembly) has to be balanced so that it meets the specified grade. With small pump impellers (up to about 250 mm diameter) this is done by removing metal either by machining or hand grinding. For larger designs, and in particular impellers with a depth:diameter ratio of more than 1:2, it is practical to add balancing weights. In these sizes it is common for the impeller to be dynamically balanced using two correction planes. Figure 7.11 shows a typical example of a seawater pump impeller in which balance is achieved using a combination of the two methods. Accurately weighted bars are welded into the nose cone of the impeller while material is machined off the inside of the hub end by mounting the impeller 'off centre' in a vertical jig boring machine. There is often a limit on how much metal can be removed – a maximum 25 per cent of the impeller shroud ring wall thickness (which is where the metal is normally removed from) is a good rule of thumb.

7.10 Pump components – clearances and fits

The correct clearances and fits are a basic but important part of a pump's fitness for purpose. The importance of obtaining the accurate dimensions, particularly on bore diameters, increases with pump size. Figure 7.13 shows typical categories of fit used for a vertical cooling water pump. These are chosen from the common mid-range toleranced fits given in ISO R286. Expect to see (as a guideline), the following categories:

- Bearing to shaft sleeve – running clearance fit
- Impeller to shaft – transition fit better than H7/k6
- Casing wear-ring to casing – locational interference fit
- Impeller to casing wear-ring – running clearance fit
- Casing section joints – transition 'spigot' fit (for tight location rather than accuracy)
- Shaft 'muff' couplings – 'sliding' clearance fit, e.g. H7/g6
- Bearing housing to bearing – locational clearance fit, e.g. H7/h6

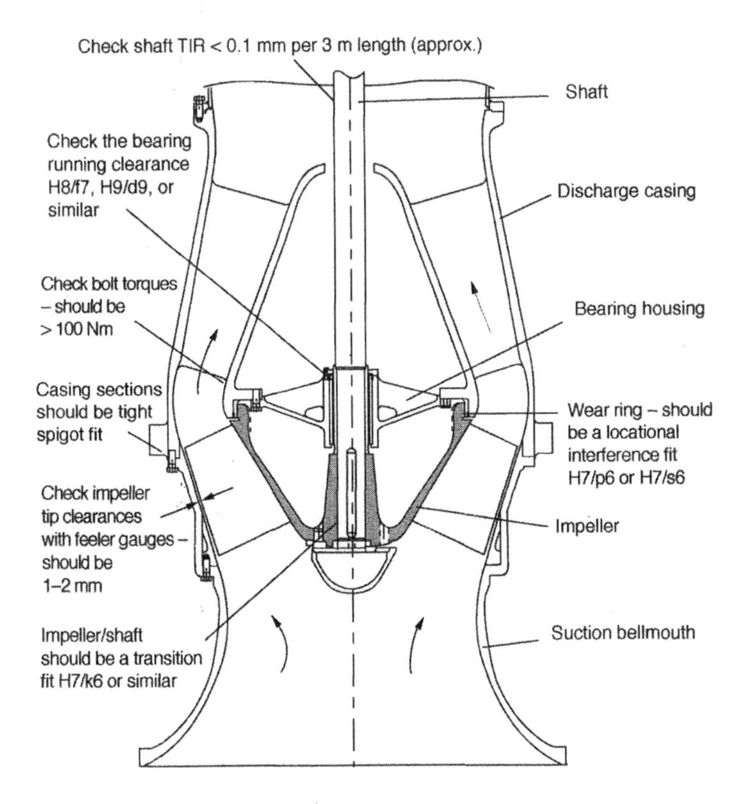
Check shaft TIR < 0.1 mm per 3 m length (approx.)
Shaft
Check the bearing running clearance H8/f7, H9/d9, or similar
Discharge casing
Check bolt torques – should be >100 Nm
Bearing housing
Casing sections should be tight spigot fit
Wear ring – should be a locational interference fit H7/p6 or H7/s6
Check impeller tip clearances with feeler gauges – should be 1–2 mm
Impeller
Impeller/shaft should be a transition fit H7/k6 or similar
Suction bellmouth

Fig. 7.13 Pump assembly checks

Table 7.3 shows some of the wide range of technical standards relevant to centrifugal pumps.

Table 7.3 Technical standards – centrifugal pumps

Standard	*Title*	*Status*
BS 5257: 1975	Specification for horizontal end-suction centrifugal pumps (16 bar).	Current
BS 5316-1: 1976, ISO 2548-1973	Specification. Acceptance tests for centrifugal, mixed-flow, and axial pumps. Class C tests.	Current
BS 5316-2: 1977, ISO 3555-1977	Specification for acceptance tests for centrifugal, mixed-flow, and axial pumps. Class B tests.	Current
BS ISO 3069: 2000	End-suction centrifugal pumps. Dimensions of cavities for mechanical seals and for soft packing.	Current
BS EN ISO 5198: 1999	Centrifugal, mixed-flow, and axial pumps. Code for hydraulic performance tests. Precision class.	Current
BS EN ISO 9905: 1998	Technical specifications for centrifugal pumps. Class I.	Current
BS EN ISO 9906: 2000	Rotodynamic pumps. Hydraulic performance acceptance tests. Grades 1 and 2.	Current
BS EN ISO 9908: 1998	Technical specifications for centrifugal pumps. Class III.	Current
BS EN 733: 1995	End-suction centrifugal pumps, rating with 10 bar with bearing bracket. Nominal duty point, main dimensions, designation system.	Current
BS EN 735: 1995	Overall dimensions of rotodynamic pumps. Tolerances.	Current
BS EN 1151: 1999	Pumps. Rotodynamic pumps. Circulation pumps having an electrical effect not exceeding 200 W for heating installations and domestic hot water installations. Requirements, testing, marking.	Current
BS EN 22858: 1993, ISO 2858: 1975	End-suction centrifugal pumps (rating 16 bar). Designation, nominal duty point, and dimensions.	Current

Table 7.3 Cont.

BS EN 23661: 1993, ISO 3661: 1977	End-suction centrifugal pumps. Baseplate and installation dimensions.	Current
BS EN 25199: 1992, ISO 5199-1986	Technical specifications for centrifugal pumps. Class II.	Current, work in hand
93/303211 DC	Firefighting pumps. Part 1. Requirements for firefighting centrifugal pumps with primer (prEN 1028-1).	Current, draft for public comment
93/303212 DC	Firefighting pumps. Part 2. Testing of firefighting centrifugal pumps with primer (prEN 1028-2).	Current, draft for public comment

CHAPTER 8

Compressors and Turbocompressors

8.1 Compressors

Compressor designs vary from those providing low-pressure delivery of a few bars up to very high-pressure applications of 300 bars. The process fluid for general industrial use is frequently air, while for some specialized process plant applications it may be gas or vapour. There are several basic compressor types, the main difference being the way in which the fluid is compressed. These are:

- *Reciprocating compressors* The most common positive displacement type for low-pressure service air. A special type with oil-free delivery is used for instrument air and similar critical applications.
- *Screw compressors* A high-speed, more precision design used for high volumes and pressures and accurate variable delivery.
- *Rotary and turbocompressors* High-volume, lower-pressure applications. These are of the dynamic displacement type and consist of rotors with vanes or meshing elements operating in a casing.

Other design types are: lobe-type (Rootes blowers), low-pressure exhausters, vacuum pumps, and various types of low-pressure fans.

Compressor performance

The main fitness-for-purpose (FFP) criterion for a compressor is its ability to deliver a specified flowrate of air or gas at the pressure required by the process system. Secondary FFP criteria are those aspects that make for correct running of the compressor; the most important one, particularly for reciprocating designs, is vibration.

The main performance-related definitions are listed in ISO 1217. They are:

- *Total pressure* p Total pressure is pressure measured at the stagnation point, i.e. there is a velocity effect added when the gas stream is brought to rest. In a test circuit, 'absolute total pressure' is measured at the compressor suction and discharge points for use in the calculations.
- *Volume flowrate* q There are three main ways of expressing this (see Fig. 8.1).
- *Free air delivery* q (FAD) This is the volume flowrate measured at compressor discharge and referred to free air (the same as atmospheric conditions). It is the definition nearly always quoted in compressor acceptance guarantees.
- *Actual flowrate* This is the volume flowrate, also measured at compressor discharge, but referred specifically to those conditions (these are total measurements) existing at the compressor inlet during the test.
- *Standard flowrate* This is nearly the same as FAD. It is the volume flowrate, measured at the discharge, but referred to a standard set of inlet conditions. A common set of standard conditions is 1.013 bar and 0 °C (273 K). A correction factor is needed to convert to FAD.
- *Specific energy requirement* This is the shaft input power required per unit of compressor volume flowrate. Power is normally an acceptance guarantee parameter.

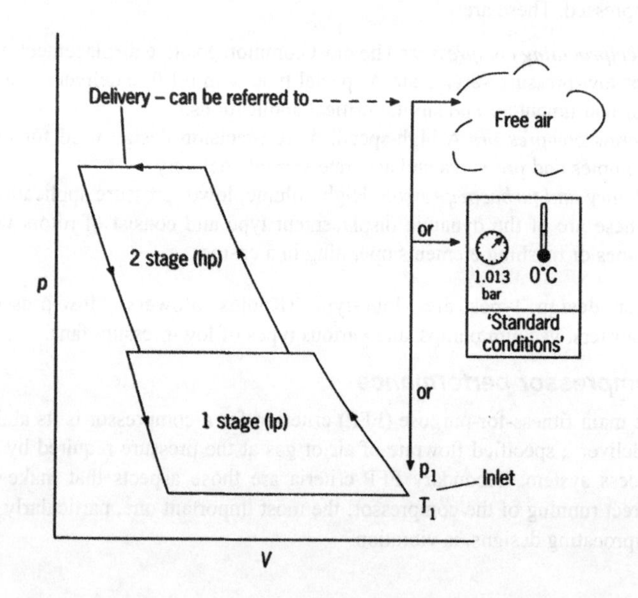

Fig. 8.1 Three ways of expressing compressor flowrate

Compressor acceptance testing

The overall objective of a compressor acceptance test is to check compliance with the specified performance guarantees, which will look something like this:

- specified inlet pressure p_1 and inlet temperature T_1; note that these are often left implicit, perhaps being described as 'ambient' conditions;
- required FAD capacity q at delivery pressure p_2;
- power consumption P at full load;
- vibration – normally specified for compressors as a velocity (using VDI 2056);
- noise – expressed as an 'A-weighted' measurement in dB(A).

Test circuits

There are several possible layouts of test circuits. The most common type for air applications is the 'open' circuit, i.e. the suction is open to atmosphere, which is representative of the way that the compressor will operate when in service. If an above-atmospheric suction pressure is required then the test circuit will be a closed loop.

The flowrate (FAD) can be measured using an orifice plate flow meter on either the suction or discharge side of the compressor. Typically, it is measured on the discharge side, with a receiver vessel interposed between the orifice and the compressor discharge (see Fig. 8.2).

The test

The performance test itself consists of the following steps:

- circuit checks;
- run the compressor until the system attains steady state conditions (up to 4 h);
- check the system parameters comply with allowable variations as in ISO 1217;
- make minor adjustments as necessary, but only those essential to maintain the planned test conditions;
- take readings at regular intervals (say 15 min) over a period of 2–4 h with the compressor running at full load;
- check again for obvious systematic errors in the recorded parameters;
- carry out functional checks of unloading equipment, relief valves, trips, and interlocks;
- perform the noise and vibration measurements;
- do the performance calculations and compare the results with the guarantee requirements (see Table 8.1).

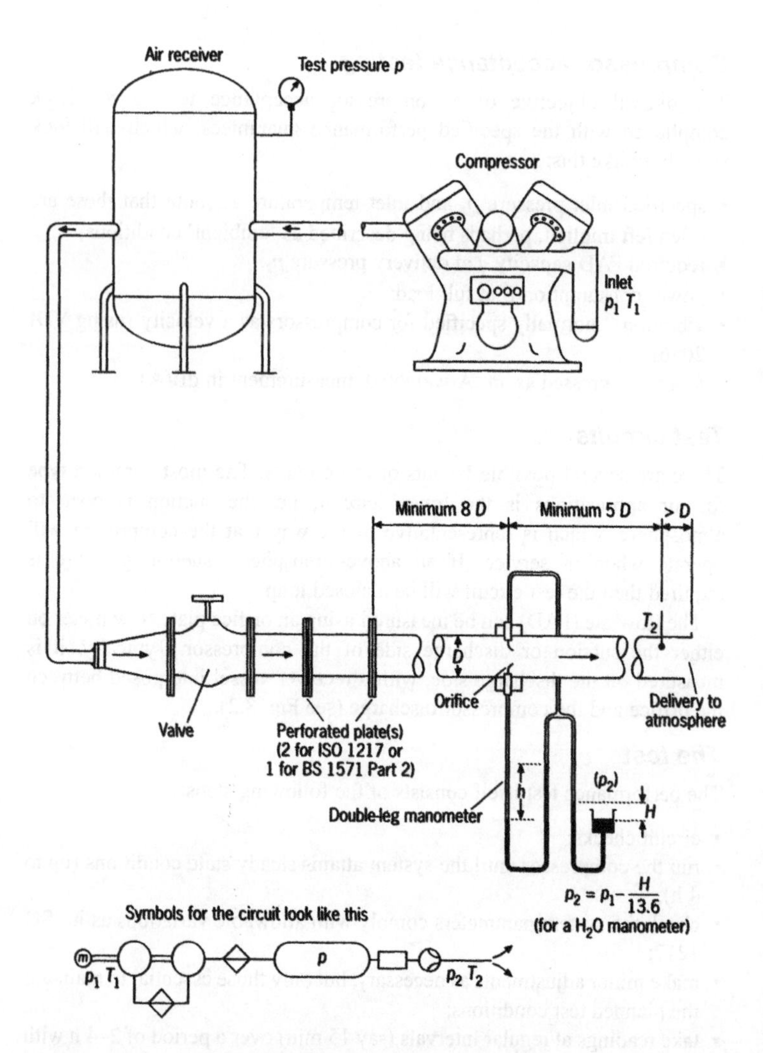

Fig. 8.2 Compressor test circuit

Table 8.1 Evaluation of performance test results

Step 1	Calculate q (FAD) by $q(FAD) = \frac{kT_1}{p_1}\sqrt{\frac{h.p_2}{T}}$	Where q = Volume flowrate h = Pressure drop across nozzle (mmH_20) K = Nozzle constant (remember the check described earlier) T_1 = Temperature (absolute K) at compressor inlet T = Temperature (absolute K) downstream of the nozzle p_1 = Pressure (absolute mmHg) at compressor inlet p_2 = Pressure (absolute mmHg) downstream of the nozzle
Step 2	This q(FAD) will be in l/s. Convert to m^2/h using $m^2/h = 1/s \times 3.6$	
NEXT	Do you need any conversion factors?	Absolute pressure – gauge reading + atmospheric pressure (check the barometer)
		REMEMBER
Step 3	If the test speed is different from the rated speed, correct the q(FAD) by q(FAD) (corrected) = q)FAD)(test) $\times \frac{\text{rated speed}}{\text{test speed}}$	This simplified correction is normally the only one you will need for a test under BS 1571 Part 2
THEN		REMEMBER
Step 4	Compare it with the q(FAD) requirements of the guarantee	There is an allowable tolerance of ± 4–6% at full load depending on the size of compressor – check with BS 1574 if in doubt
Step 5	Check power consumption kW	Normally measured using two wattmeters
	Any correction if necessary Power (corrected) = Power (test) $\times \frac{\text{rated speed}}{\text{test speed}}$	
	If specifically required by the guarantee express power consumption in 'specific energy' terms by: Specific energy = $\frac{\text{energy consumption}}{q(FAD)}$	Watch the units: a normal unit is kW h/l

Compressor specifications and standards

The most commonly used test standard is:

- ISO 1217: 1986 *Methods for acceptance testing* (identical to BS 1571 Part 1).

This gives comprehensive testing specifications and arrangements for the major compressor types. It is particularly suitable for testing an unproven or 'special' compressor design.

Other relevant standards are shown in Table 8.2.

Table 8.2 Technical standards – compressors

Standard	*Title*	*Status*
BS 1553-3: 1950	Graphical symbols for general engineering. Graphical symbols for compressing plant.	Current
BS 1571-2: 1975	Specification for testing of positive displacement compressors and exhausters. Methods for simplified acceptance testing for air compressors and exhausters.	Current, confirmed
BS 1586: 1982	Methods for performance testing and presentation of performance data for refrigerant condensing units.	Current
BS 1608: 1990	Specification for electrically driven refrigerant condensing units.	Current
BS 3122-1: 1990, ISO 917: 1989	Refrigerant compressors. Methods of test for performance.	Current
BS 6244: 1982, ISO 5388-1981	Code of practice for stationary air compressors.	Current
BS 7316: 1990	Specification for design and construction of screw and related type compressors for the process industry.	Current, proposed for withdrawal
BS 7321: 1990, ISO 8011: 1988	Specification for design and construction of turbo-type compressors for the process industry.	Current, proposed for withdrawal
BS 7322: 1990, ISO 8012: 1988	Specification for design and construction of reciprocating-type compressors for the process industry.	Current, proposed for withdrawal

Table 8.2 Cont.

BS 7854-3: 1998, ISO 10816-3: 1998	Mechanical vibration. Evaluation of machine vibration by measurements on non-rotating parts. Industrial machines with nominal power above 15 kW and nominal speeds between 120 r/min and 15 000 r/min when measured in situ.	Current
BS ISO 1217: 1996	Displacement compressors. Acceptance tests.	Current
BS EN 255-1: 1997	Air conditioners, liquid chilling packages, and heat pumps with electrically driven compressors. Heating mode. Terms, definitions, and designations.	Current
BS EN 1012-1: 1997	Compressors and vacuum pumps. Safety requirements. Compressors.	Current
BS EN 1012-2: 1997	Compressors and vacuum pumps. Safety requirements. Vacuum pumps.	Current
BS EN 12583: 2000	Gas supply systems. Compressor stations. Functional requirements.	Current
BS EN 12900: 1999	Refrigerant compressors. Rating conditions, tolerances and presentation of manufacturer's performance data.	Current
95/710815 DC	Measurement of noise emission from compressors and vacuum pumps (engineering method) (prEN 12076).	Current, draft for public comment
95/715797 DC	Petroleum and natural gas industries. Rotary-type positive displacement compressors. Part 2. Packaged air compressors. (Joint TC 118-TC 67/SC 6) (ISO/DIS 10440-2.)	Current, draft for public comment
95/715798 DC	Petroleum and natural gas industries. Rotary-type positive displacement compressors. Part 1. Process compressors. (Joint TC 188/TC 67/SC 6) (ISO/DIS 10440-1.)	Current, draft for public comment
96/702049 DC	prEN ISO 917. Testing of refrigerant compressors.	Current, draft for public comment
96/702050 DC	prEN ISO 9309. Refrigerant compressors. Presentation of performance data.	Current, draft for public comment

Table 8.2 Cont.

96/706683 DC	Reciprocating compressors for the petroleum and natural gas industries (ISO/DIS 13707).	Current, draft for public comment
96/708979 DC	Centrifugal compressors for general refinery service in the petroleum and natural gas industries (ISO/DIS 10439).	Current, draft for public comment
96/716066 DC	Petroleum and natural gas industries. Packaged, integrally geared centrifugal air compressors for general refinery service (ISO/DIS 10442).	Current, draft for public comment
97/700630 DC	Refrigerating systems and heat pumps. Safety and environmental requirements. Refrigerant compressors (prEN 12693).	Current, draft for public comment
99/716305 DC	prEN 13771-1. Refrigerant compressors and condensing units for refrigeration. Performance testing and test methods. Part 1. Refrigerant compressors.	Current, draft for public comment
00/706276 DC	ISO/DIS 5389. Turbocompressors. Performance test code.	Current, draft for public comment
BS 1571: Part 1: 1975, ISO 1217-1974	Specification for testing of positive displacement compressors and exhausters. Acceptance tests.	Withdrawn, revised

8.2 Turbocompressors

For very large volume throughputs of gas (generally air) at low pressure, a centrifugal dynamic displacement machine is used, known generically as a 'turbocompressor'. Turbocompressors may have single or multiple pressure stages, depending on the delivery pressure and volume required. Although they work using the same thermodynamic and fluid mechanics 'rules' as reciprocating or screw compressors, they are very different mechanically, the main reason being because of their high rotational speed. Turbocompressors are fast and potentially dangerous machines – a typical 300–600 kW machine can have rotational speed of up to 25 000 r/min, giving impeller tip speeds of around 500 m/s. Typical uses are for large-scale ventilation of enclosed spaces and for aeration of fluid-bed based chemical processes such as gas desulphurization or effluent treatment plant.

Performance and guarantees

Turbocompressor guarantees are based on the typical performance characteristic shape shown in Fig. 8.3. This consists of a series of operating curves (representing different vane settings) plotted within a set of pressure–volume axes. There is an upper operating limit at the top of the pressure range. The machine will not operate properly above this point, known as the 'surge line'. The characteristic shown is typical for a single-stage turbocompressor – the range and gradients will be different for a multi-stage machine. Another guarantee criterion is noise. Turbocompressors are noisy machines: a large unit of 600 kW with a 500 m/s vane tip speed will produce in excess of 98 dB(A). This is above the safety level for human hearing, so most turbocompressors are fitted with an acoustic enclosure.

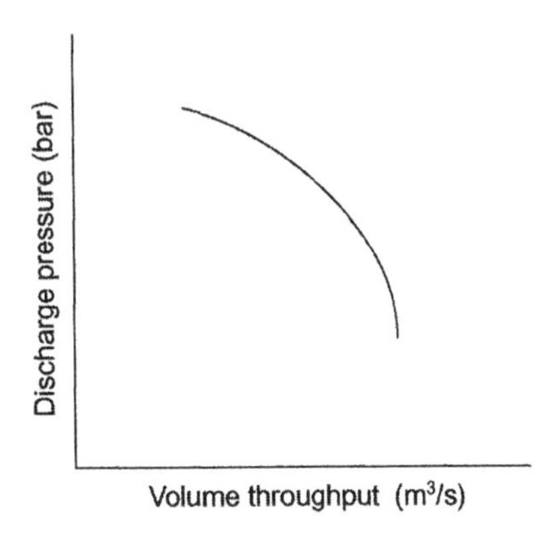

Fig. 8.3 Turbocompressor – the basic performance characteristic

Mechanical arrangement

Figure 8.4 shows the general mechanical arrangement of a single-stage turbocompressor. The mechanical design features that differ from those found on a standard air compressor are:

- a step-up gearbox;
- tilting pad thrust bearings;
- inlet guide vanes (IGVs) and/or variable diffuser vanes (VDVs);
- highly rated bearings (100–300 000 h rated life);
- axial alignment of gear wheels accurate to about 10 μm;
- back-up lubricating oil system;
- precision 'Hirth toothed' shaft coupling.

The turbocompressor casings can be fabricated or cast. Special sprung resilient mountings are used to minimize the transmission of structural vibration.

The fluid characteristics of turbocompressors work on the same concept as other rotodynamic machinery: the principle of changing dynamic energy into static energy, i.e speed is 'converted' into pressure. Downstream diffusers and the specially shaped spiral casing are used to help optimize the flow regime. Air delivery pressure is limited by the position of the surge line on the characteristic, so guarantees are normally quoted in terms of a delivery pressure in 'water gauge' (metres of H_2O) at a rated continuous volume throughput.

Compared to other compressor types, the delivery pressure is low, normally a few bars. The absolute maximum for centrifugal machines is 18 bar, although few practical machines go this high, except for perhaps sparge-air applications in some specialist chemical processes. Inlet air conditions are quoted as an absolute temperature and pressure, with the addition of relative humidity. This makes the performance calculations for turbocompressors more complicated than the simplified free air delivery (FAD) method used for normal compressors, so specialist technical standards are needed. Both single- and multi-stage machines are specified with a minimum 'turndown' ratio (normally 40–60 per cent) set by the process system that the turbocompressor supplies. Machine efficiencies are high – near 90 per cent.

Specifications and standards

There are at least three different sets of standards in common use: VDI and BS of European origin and the ASME PTC from the USA. They use similar principles and methodologies, although differing slightly in some areas of detail.

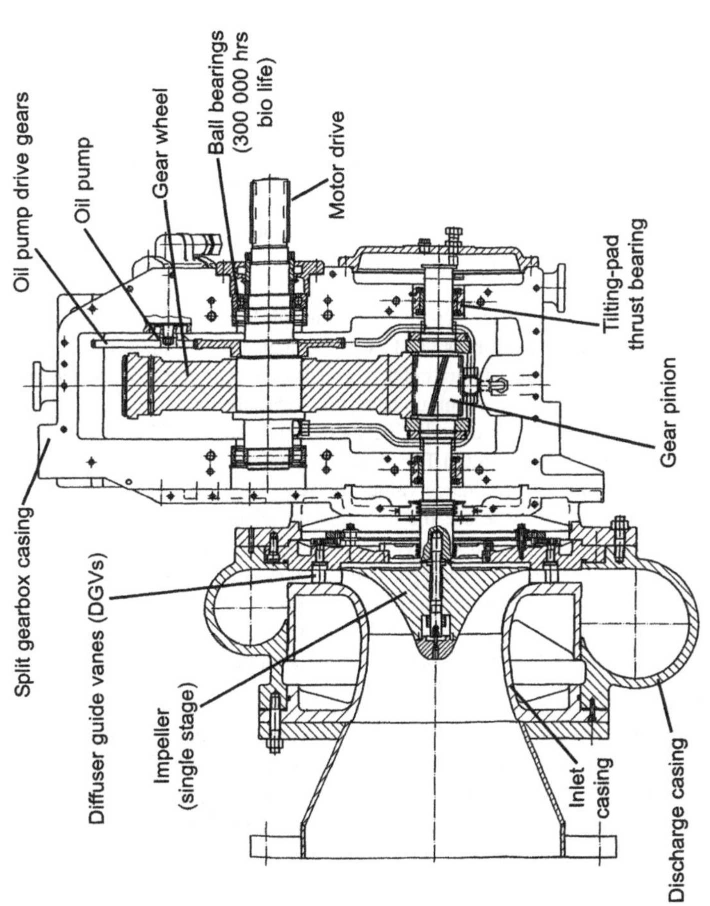

Fig. 8.4 Single-stage turbocompressor – general mechanical arrangement

Acceptance and performance tests

VDI 2045 (Table 8.3) is the most comprehensive technical standard available on the subject of turbocompressor performance tests. VDI 2045 is unlike the comparable BS or ASME PTC standard in that it relates to both turbocompressors and positive displacement compressors. There are no significant technical contradictions raised by this approach; it just means that the standard contains some duplication of sections, to cater for the two different machine types. Table 8.3 is a guide to finding the most important pieces of information in the standard.

Part 1 concentrates on guarantee testing and the principles of the conversion and comparison of test results. Important parts are:

- *Emphasis on performance guarantees*: it concentrates on the three main guarantee parameters: fluid volume throughput, discharge pressure, and power consumption.
- *Measurement uncertainties*: the same philosophy of measurement uncertainties and errors is used as in the standard for pump testing, ISO 3555. Each measured parameter is given a level of accuracy, depending on the instrument or technique used, and these accumulate into an overall uncertainty level. (It can also be thought of as a confidence level that applies to the test results.) The standard itemizes these uncertainties, and suggests suitable percentage values to use.
- *Test deviations*: these are the amounts by which the various measured parameters are allowed to deviate during a performance test and still be considered acceptable readings. The standard gives recommended levels.
- *Referenced standards*: one of the strong points of VDI 2045 is the clear way it cross-references other related standards that apply to the test. Most of these are to do with the hardware and layout of the test rig.

ASME PTC-10 is the most common code used in the USA and in the offshore and process industries. As with the other standards, it concentrates on performance aspects rather than the mechanics of turbocompressor machines. The unique thing about PTC-10 is that it caters for three classes of turbocompressor tests. Class I is used when the test gas and arrangement are the same as the machine will see in service. Class II and III involve a degree of performance prediction, i.e. where the test gas is different, normally for reasons of safety. Turbocompressors for explosive gas service are normally tested on air, under the provisions of PTC-10 Class II or III. The difference between Class II and III is only in the way of processing the test results, depending on the level of 'real gas' assumptions that are used during the calculations. Table 8.4 shows other standards referenced by PTC-10.

An old, well-established standard dedicated to turbocompressors is BS 2009. Sections 1 and 2, which are short, cover temperature and pressure measurement techniques and symbols. Section 3 is about the correction of acceptance test results to guarantee conditions. There are no separate test classes, as in PTC-10. Diagrams of acceptable test layouts are shown in the appendix of the standard.

Table 8.3 Important information in VDI 2045

VDI 2045: 1993: *Acceptance and performance tests on turbocompressors and displacement compressors*

Part 1: *Test procedure and comparison with guaranteed values*

Part 2: *Theory and examples*

Referenced standards

- VDI 2045 *Suction line inlet diaphragm*
- DIN 1952
- VDI 2059: Part 3 *Shaft vibration of industrial turbosets – measurement and evaluation*
- VDI 2056 *Criteria for assessing mechanical vibration of machines*. This standard deals with 'housing' vibration and is similar to ISO 2372/BS 4675

BS 2009: *Code for acceptance tests for turbo-type compressors and exhausters*

Referenced standards

- BS 1571: Part 2: 1984 *Method for simplified acceptance testing for air compressors and exhausters*. This replaces BS 726
- BS 1042: Part 1 (various sections): *Pressure differential devices*
- BS 848: Part 1: 1980 *Fans for general purposes – methods of testing performance*

VDI 2045 Part 1

Subject	*Section*
The objective of guarantees	1.3.3
'Type test' acceptance	1.3.6
List of symbols and indices	2
The principle of measurement uncertainty	3.1.1
Important measurement guidelines (cross-references)	3.1.2
Fluid volumetric/mass flow guidelines	3.7.1
Power measurement	3.9.2
Energy balance assumptions	3.9.4

Table 8.3 Cont.

Performance test	: preparation	4.1
	: general requirements	4.2
	: allowable deviations	4.2.2 and 5.3. Tables 5 and 6
Measurement uncertainties (test results)		4.4
Conversion of test results to guarantee conditions		5.3, Figs 3 and 7, Table 5
Typical performance curves		5.3.5, 6.1, 6.2.2.2, 6.2.3, 6.2.4
Testing reporting and documentation requirements		7

VDI 2045 Part 2

Subject	***Section***
Reference boundaries	2.3 and Fig. 3
Performance curves	2.9.1
Guide vane and surge effects	2.9.1
Specimen test results	3.3.2
Conversion to guarantee conditions	3.3.1

Table 8.4 Turbocompressors – ASME PTC-10 referenced standards

ASME PTC-10: 1986: *Compressors and exhausters*

Referenced standards

- PTC-1 *General instructions*
- PTC-19.3 *Instruments for temperature measurements*
- PTC-19.5 *Flow measurement*
- PTC-19.3 *Measurement of rotational speed*
- PTC-19.7 *Measurement of power*
- ASA 50.4 *Motor efficiency measurement* IEEE
- PTC-19.6 *Electrical measurement in power circuits*
- PTC-19.5 *Flow coefficients for orifices*

Vibration measurement

An important standard is VDI 2059. It is relevant for several types of turbomachinery, but its principal use is for turbocompressors, mainly because of their higher speed – up to 30 000 r/min, compared to a practical maximum of 5000–6000 for other types of turbomachinery. VDI 2059 relies on four fundamental principles.

- VDI 2059 is about vibration of the shaft *relative to* its journal bearings. This is unrelated to the simpler types of 'housing' vibration covered by VDI 2056/BS 4675/ISO 2372.
- Vibration is sensed by non-contacting probes which look at highly polished areas of the shaft. They are located in two perpendicular (x and y) planes.
- The measured parameter is shaft displacements measured in microns, not velocity as used for 'housing' vibration. There are two aspects to this: the absolute level of shaft displacement measurement during a test run, and the amount by which the displacement changes as the test progresses. These are referred to by the standard as criterion I and II, respectively.
- The concept of VDI 2059 shaft vibration is that it is not sinusoidal, or in any way sine-related. This means that it is necessary to look at the shape of the shaft's path (known as its 'orbit') to describe fully the vibration, before deciding a level that is deemed acceptable for turbocompressor operation.

These principles are reflected in the way that shaft vibration levels are assessed under VDI 2059. The important points are shown in Fig. 8.5. Note the annotations that show how the maximum recorded displacement in the x and y planes are resolved to give a resultant value. This standard includes guidance on acceptance levels. There are three classes designated: A (the most stringent), B and C for absolute (the so-called criteria I value) displacement levels, and a single acceptance level for the amount by which displacement levels change during the test.

The performance test

The main aspects of interest are:

- *the discharge pressure/volume characteristic*: the objective is to demonstrate whether the machine will reach the specified volume throughput of gas at the required discharge pressure, without surging;
- *the turn-down ratio* to demonstrate that the flow modulation vanes can be adjusted sufficiently to reach the minimum rated throughput;
- *mechanical integrity*: vibration levels must be controlled on such high-speed machines, and the turbocompressor needs to be able to run continuously without any significant wear or deterioration;

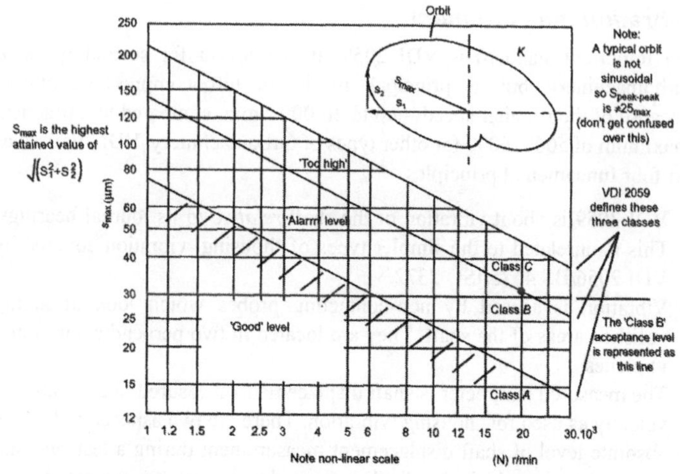

Absolute displacement (Criterion I) acceptance levels

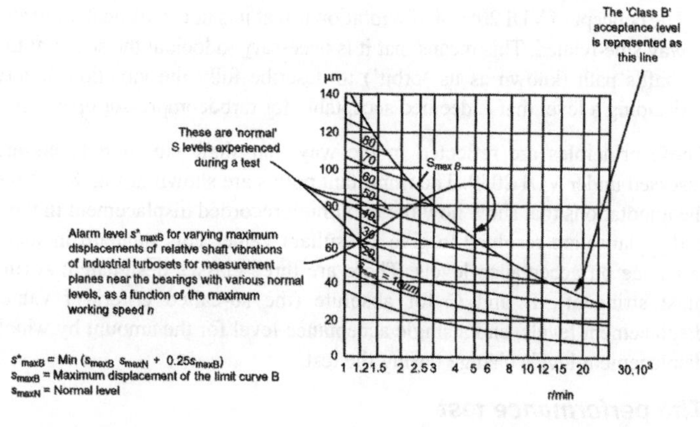

Displacement change (Criterion II) acceptance levels

Fig. 8.5 Shaft vibration – the essential points of VDI 2059

- *noise level* normally with an acoustic enclosure
- *efficiency or power consumption*, depending on the exact form of specification used.

Table 8.5 shows a typical set of performance guarantees expressed for such a low-discharge pressure air sparging turbocompressor.

Table 8.5 Low-pressure turbocompressor: typical performance guarantees

Single-stage turbocompressor, aircooled with variable diffuser guide vanes. Motor speed 3000 r/min with single helical step-up gearbox to impeller speed of 22 900 r/min. Skid-mounted with integral oil tank and acoustic sound enclosure.		
Medium	Atmospheric air	The inlet conditions
Inlet temperature	28 °C	
Inlet pressure	1013 mBar (absolute)	
Relative humidity	100%	
Gas constant	288.9 Nm/kg/K	Assumptions
Isentropic index	1.4	
Volume flow (measured at suction)	Minimum 5000 m^3/h ($83.3\ m^3$/min) Maximum 11 000 m^3/h ($183.3\ m^3$/min)	Performance requirements
Discharge pressure	2 bar absolute without surging (approx. 10 m 'water gauge')	
Power consumption	320 kW (+3% tolerance)	
Noise	85 dB(A) 1 m from enclosure	
Performance test to VDI 2045 with vibration assessment to VDI 2059 Class B		Test standard

The test arrangement

The most common type of turbocompressor performance test is carried out 'open-circuit' using air, hence using atmospheric air as a near-approximation to the specified suction conditions. The test circuit arrangement for this is shown in VDI 2045 and the other relevant technical standards and is summarized in Figs 8.6 and 8.7.

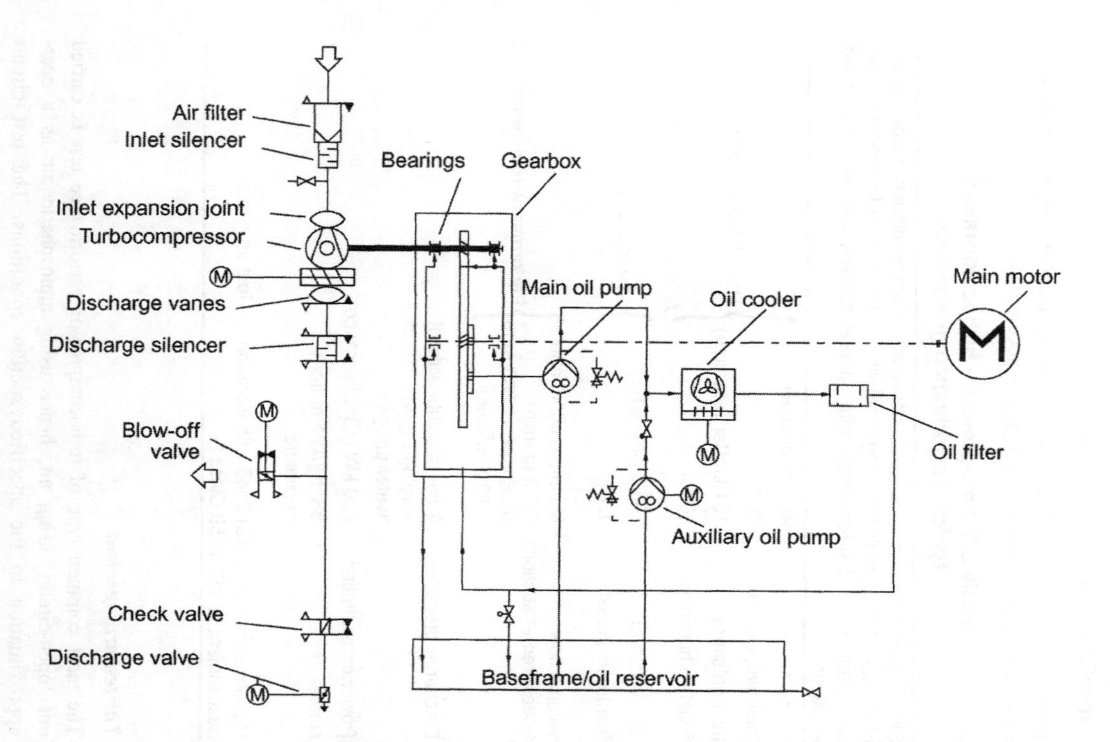

Fig. 8.6 Turbocompressor performance test – typical test circuit

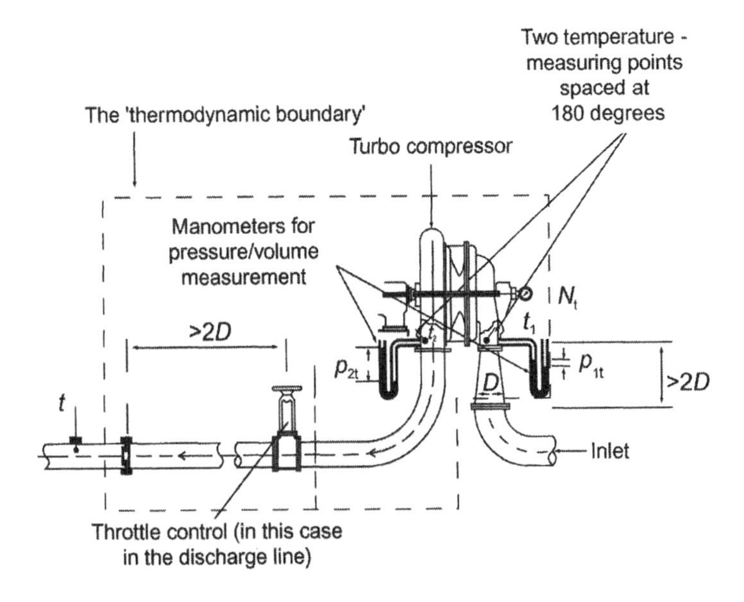

Fig. 8.7 Turbocompressor performance test – flow measurement

Flowrates are measured by four tappings, equispaced around the circumference of the pipe and interconnected by a loop. This gives an accurate 'averaged' pressure reading to feed to the water gauge manometer.

The performance test routine

Figure 8.8 shows typical performance test results. Note how the maximum throughput, minimum (turn-down) point, and power consumption at full load all meet the guarantee points as they are shown on the graph.

Vibration measurements

VDI 2059 uses the concept of relative shaft vibration as the acceptance parameter for turbocompressor vibration. The non-contacting probes used to sense this vibration are a permanent fixture, threaded into holes extending through the bearing housing and sensing from a highly polished area of the gear shaft. Figure 8.9 shows the typical location of the vibration sensors. Both sensing positions are located on the high-speed pinion shaft, and measurement is recorded in two perpendicular planes (x and y) at each location. The measured parameter is vibration displacement s (μm).

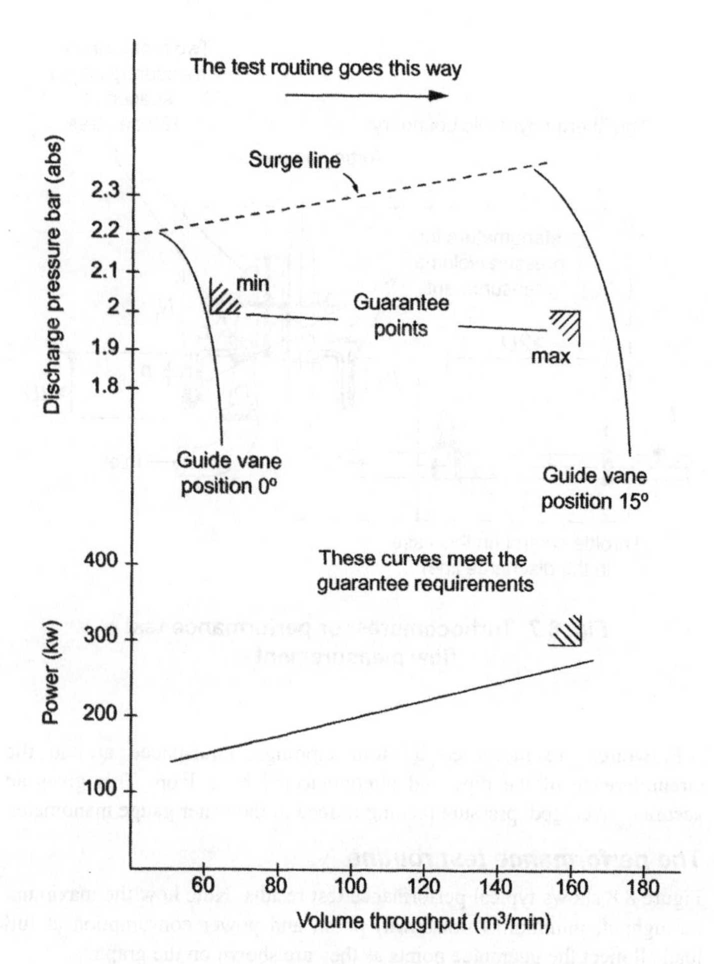

Fig. 8.8 Turbocompressor – typical 'performance map' test results

Location of axial position sensor (if fitted)

Note: remember that these are relative shaft vibrations to VDI 2059, not 'housing' vibrations as in VDI 2056

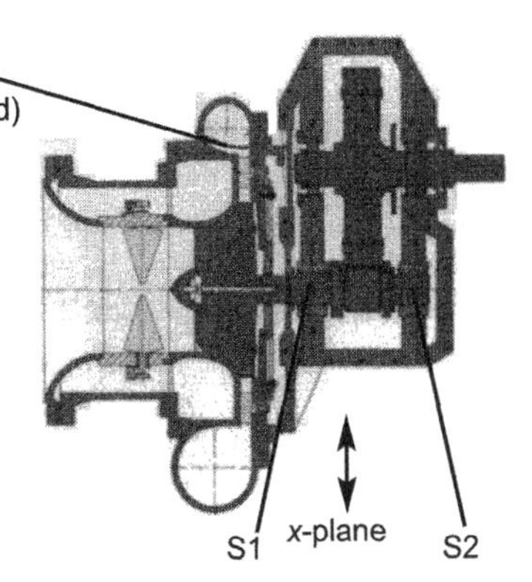

You can ignore transient vibrations at low frequencies

Shaft vibration sensors S1 and S2 both monitor the (high speed) pinion shaft

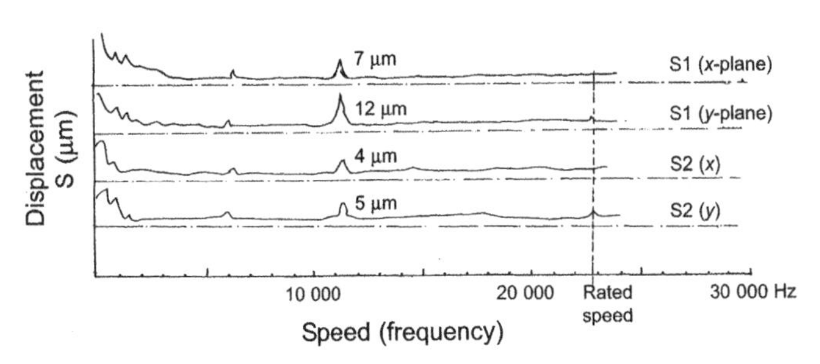

To check if the results are acceptable:

$S_{max} = \sqrt{(\hat{S}_x^2 + \hat{S}_y^2)} = \sqrt{(7^2 + 12^2)} = 13.9\mu m$

Compare this to the acceptable level $\hat{S}_B$ from Fig. 8.5 = 30 μm, so the results are acceptable to VDI 2059 Class B.

Checking the displacement change from VDI 2059, (and Fig. 8.5)

$\hat{S}_B$ = smaller of $\hat{S}_B$ or ($\hat{S}_N$ + 025 $\hat{S}_B$) ie 30 μm or (13.9 + 30/4) = 21 μm

So $\hat{S}_B^*$ = 21 μm: the greatest 'change' from the above results is (12 – 4) = 8 μm, hence the results are acceptable.

Fig. 8.9 Turbocompressor vibration tests

Displacement measurements are plotted against frequency scanning from 0 Hz up to a frequency corresponding to rotational speed of the pinion shaft – 22 900 r/min in this example. Figure 8.9 shows typical results – the various 'transient' displacements at very low frequencies can safely be ignored. The largest displacements will show as prominent 'peaks' on the trace, normally close to a fraction of the rotational frequency. It is unusual for both sensing positions on the pinion shaft to show the same displacement values: the sensor nearer the impeller will generally show the highest reading (as in Fig. 8.9). This is caused by 'imposed vibration' from the impeller, which experiences various hydrodynamic instabilities, exaggerated by its high rotational speed.

Under VDI 2059 there are two 'acceptance criteria': the maximum displacement level S_{max} (the so-called 'criterion I') and the acceptable 'displacement change' (criterion II). Figure 8.5 shows how to calculate these. Note how some simple assumptions have to be made as to what is considered the 'nominal' maximum displacement.

Noise measurement

Turbocompressors are inherently noisy machines, up to about 98 dB(A) for the highest tip speed versions, so they are nearly always fitted inside an acoustic enclosure to reduce the noise to manageable levels. A guarantee figure of 85 dB(A) outside the enclosure is normal.

The principles of noise measurement are broadly the same as used for gas turbines, i.e. the measurement of the A-weighted average at a distance of 1 m from the turbocompressor 'reference surface'.

CHAPTER 9

Prime Movers

9.1 Steam turbines

Steam turbines are complex items of rotating equipment. They can be very large, up to 1500 MW capacity with LP rotor diameters of several metres, introducing a variety of technical challenges related to large components and heavy material sections. Some typical design criteria that have to be overcome are:

- *High superheat temperatures and pressures*, with the corresponding high specification material choices.
- *Thicker material 'ruling' sections in the casing parts.* This attracts a number of particular material defects more likely to occur in thick, cast sections.
- *Longer unsupported rotor lengths.* This gives a greater tendency for bending and subsequent vibration, particularly on single-shaft machines.
- *Larger diameters*, particularly of the LP rotors, in which most of the stress on a blade is caused by centrifugal force rather than steam load. Higher stresses mean a greater sensitivity to defect size, requiring more searching NDT techniques on the rotating components.

Operating systems

Steam turbines incorporate several complex operating systems.

Lubricating oil (LO) system

Figure 9.1 shows a basic schematic diagram of a steam turbine LO system. LO pressure is maintained in the bearing galleries by means of pumps and a constant-pressure valve. Tube- or plate-type heat exchangers coupled with an automatic temperature control valve regulate the temperature. The LO

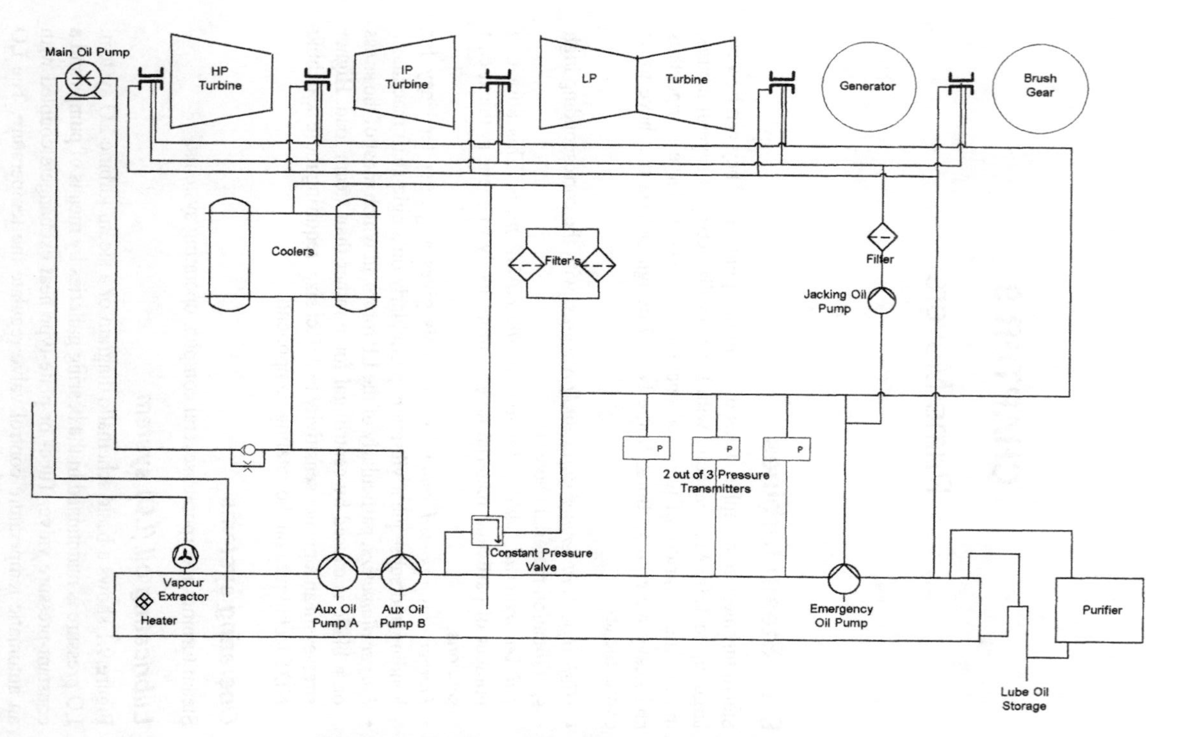

Fig. 9.1 Steam turbine LO system – schematic

drain tank underneath the turbine is designed with sufficient volume and residence time to allow the oil to de-aerate before being pumped back to the bearings. This tank is maintained at a slight vacuum by a vapour extraction fan that exhausts to atmosphere. A gear pump driven off the turbine shaft provides design LO flow at shaft speeds greater than about 80 per cent of full speed; at lower speeds, the flow is supplemented by electric pumps. A smaller capacity (approximately 60 per cent flow) back-up electric (usually DC battery-operated) pump is provided to supplement flow during system power failures. The steam turbine bearings are fed via individual oil supply lines fitted with orifice plates. On discharge from the bearings, the oil drains into the bearing pedestals through sight glasses. Temperature and pressure supervision is used to monitor running conditions.

Jacking oil

Jacking oil is used to pressurize the bearings and thereby reduce the friction coefficient between the turbine shaft and the bearings during start-up and shut-down of the turbine. Pressure is supplied by a separate positive displacement (normally a variable displacement swash-plate piston-type) jacking oil pump. The pump cuts in and out automatically when the shaft reaches pre-set rotational speeds. Figure 9.2 shows a schematic arrangement of a typical steam turbine jacking oil system.

Hydraulic system

Most steam turbine designs are fitted with a hydraulic oil system that operates the various steam admission and control valves. The system comprises triple-rotor positive displacement screw pumps supplying through a duplex filter/regeneration and a pilot-operated constant-pressure valve arrangement. An in-line accumulator may be used to provide a pressure 'reservoir' in the system. The system is normally totally separate from the turbine-lubricating oil, and uses a special grade of hydraulic fluid operating at pressures up to about 40 bar g.

The hydraulic system is used to power the turbine safety and protection system (TSPS). This is an electronically operated system that operates the steam inlet 'intercept' valves via electrohydraulic transducers. Figure 9.3 shows a simplified schematic. The entire system works on a fail-safe principle, i.e. the hydraulic pressure acts to keep the steam valves open. The trip system uses 2-out-of-3 channel logic in which operation of two trip-signal sensors is sufficient to depressurize the system and thereby trip the turbine. The trip functions are restricted during normal transient start-up and shut-down sequences of the turbine in order to avoid spurious trips.

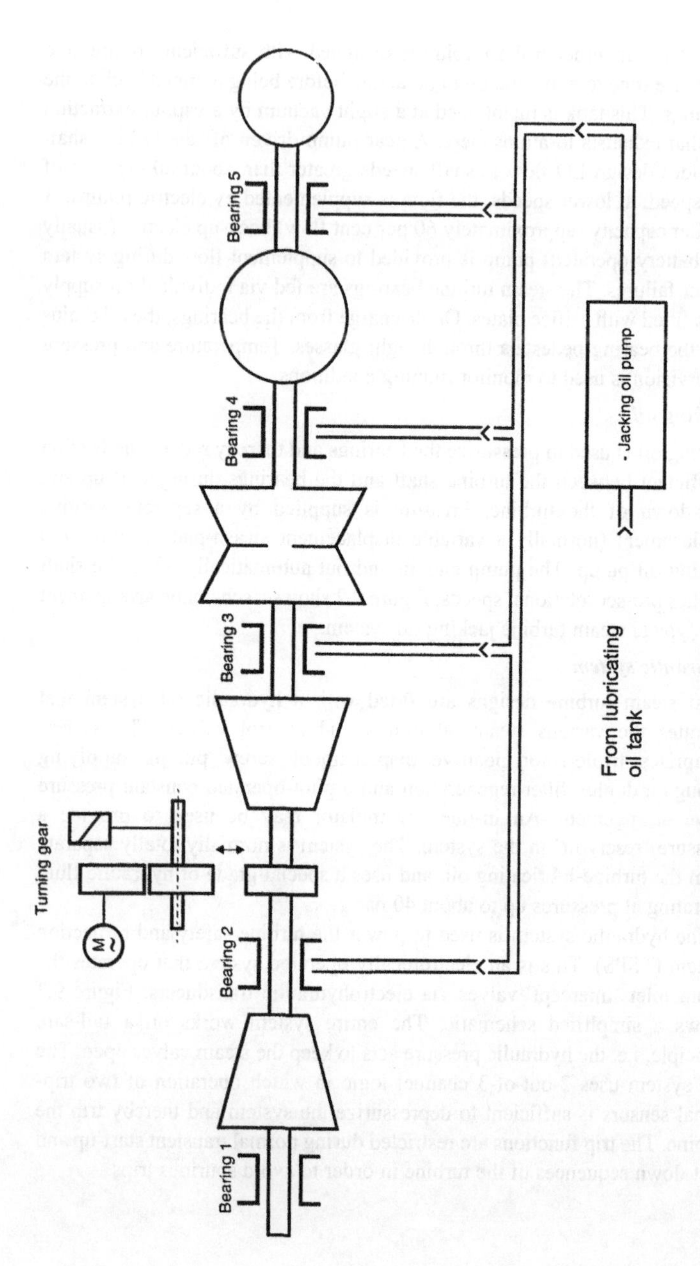

Fig. 9.2 Steam turbine jacking oil system – schematic

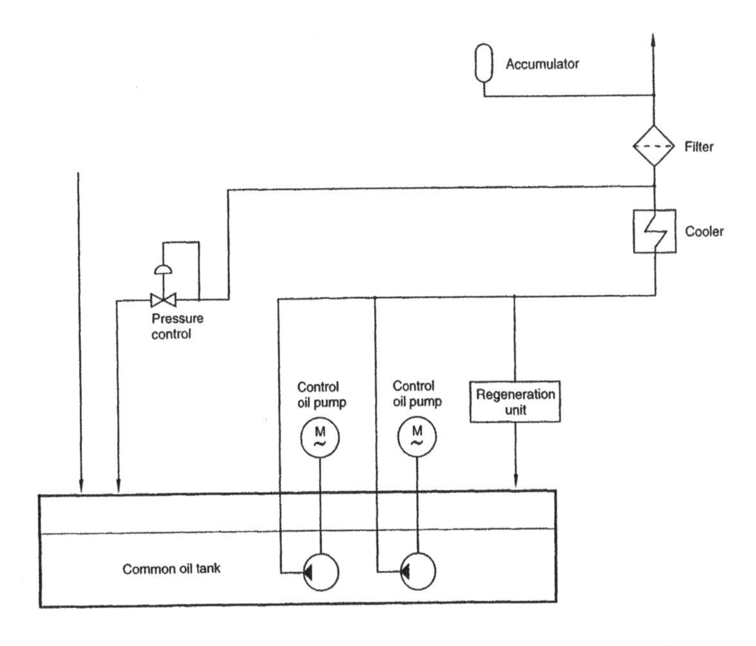

Fig. 9.3 Steam turbine hydraulic oil system – schematic

Gland steam system

All steam turbines are fitted with a gland steam system (Fig. 9.4) that stops steam leakage along the turbine shafts and prevents air being drawn into low-pressure areas, destroying the vacuum. The normal method used is non-contact labyrinth seals (see Fig. 9.5). The quality of the gland steam under all operating conditions of the turbine is controlled by an admission valve. Excess superheat temperature is reduced by means of water sprays or a similar desuperheating arrangement. Under conditions of high turbine load, excess gland steam is routed to the condenser by an automatic dump valve.

Vacuum breaker

Steam turbines are fitted with a 'vacuum breaking', electrically actuated butterfly valve that opens to allow air to enter the condenser during the run-down period when the turbine rotor is coasting to a stop. The admission of air destroys the vacuum and provides a resistance to the rotor, thereby stopping it more quickly and avoiding extended periods of operation at

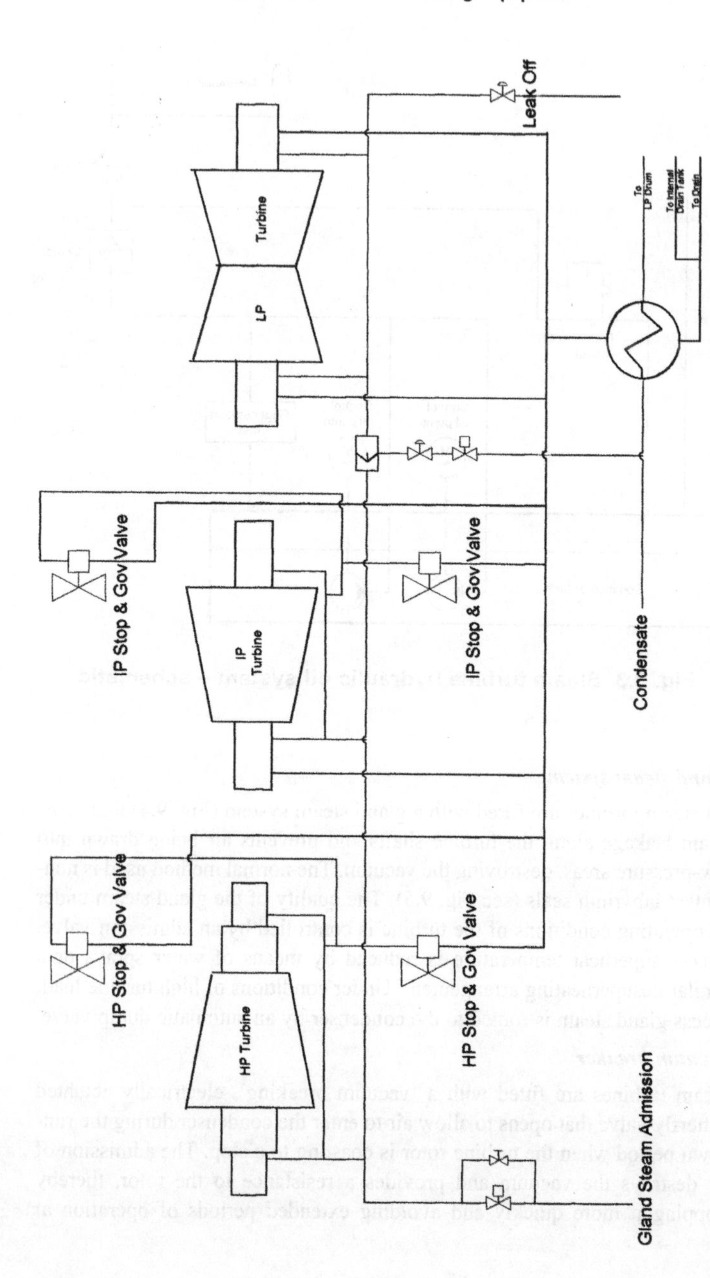

Fig. 9.4 Steam turbine gland steam system – schematic

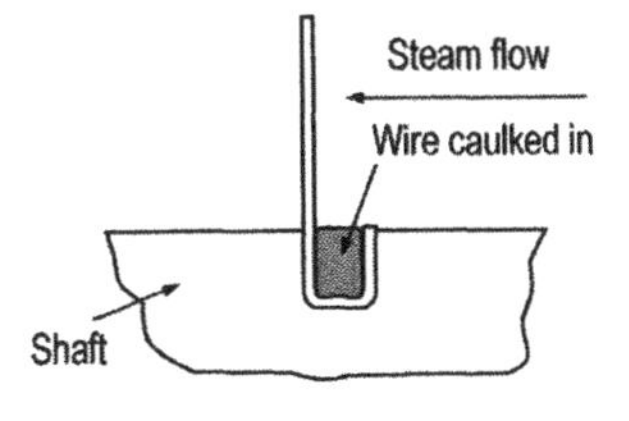

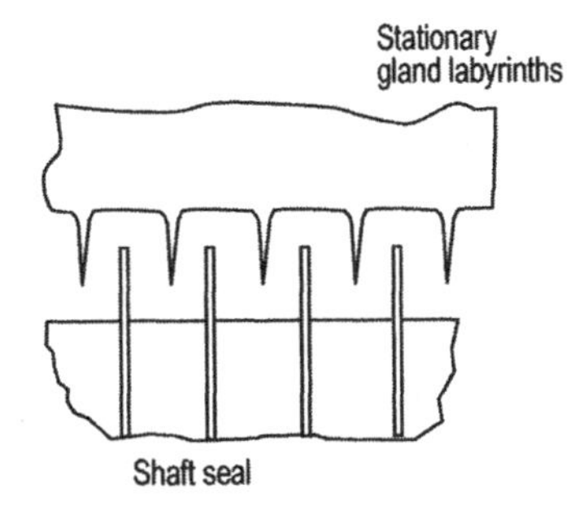

Fig. 9.5 Steam turbine labyrinth seals

critical speeds, which would cause excessive vibration and resultant high rotor stresses and bearing wear. The vacuum breaking valve opens progressively so that full atmospheric pressure is not restored in the turbine casing until shaft speed has fallen to less than about 50 per cent of rated speed. This avoids excessive stresses in the lower pressure stages of the turbine blades.

Turbine drains system

Turbine casing drain valves are installed to drain the casing during periods of start-up and transient operation, thereby helping to minimize damage from water hammer and excessive thermal stresses. Separate drain lines are used to drain condensate from specific areas during, for example, start-up. Drain valves are normally pneumatically controlled and are divided into external drains, which drain to an external atmospheric vessel and operate when the turbine is at standstill, and internal drains, which drain to an integral flash box, and operate only when the rotor is moving. Automatic drains are normally set to close when the turbine has reached approximately 15–20 per cent of full load.

Supervision systems

The turbine supervision consists of a number of electronic systems that monitor the following parameters (see Fig. 9.6):

- turbine rotational speed;
- shaft axial position relative to the casing;
- bearing housing and shaft vibration;
- absolute and differential expansions;
- rotor eccentricity.

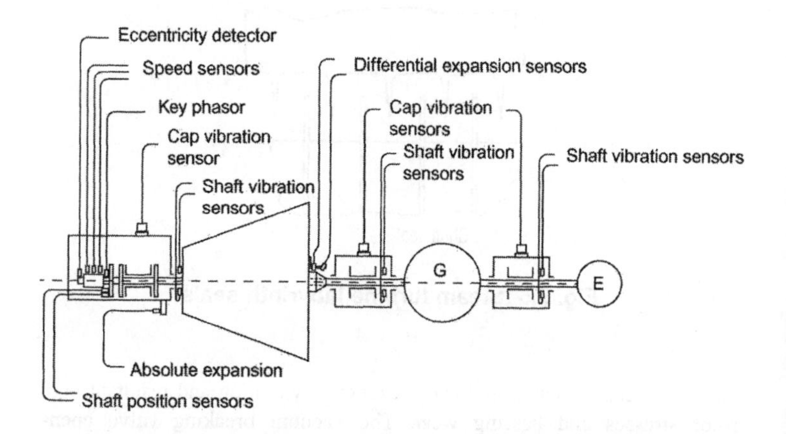

Fig. 9.6 Steam turbine supervision system – schematic

A reference 'zero' position of the rotor is fixed using a 'key phasor' position sensor. Rotational speed is measured by three non-contact probes. Axial displacement of the rotor is measured by inductive sensors located in the thrust (axial) bearing housing. Housing and shaft vibration is sensed using the principles shown in Chapter 4 and referred to by ISO 1940 or API standard limits. Figure 9.7 shows a typical monitoring arrangement.

'Absolute expansion' in steam turbines is a measure of sudden variations in expansion that do not correspond to thermal transients taking place at the time. Figure 9.8 shows the method of absolute and differential expansion measurements at the bearing pedestals.

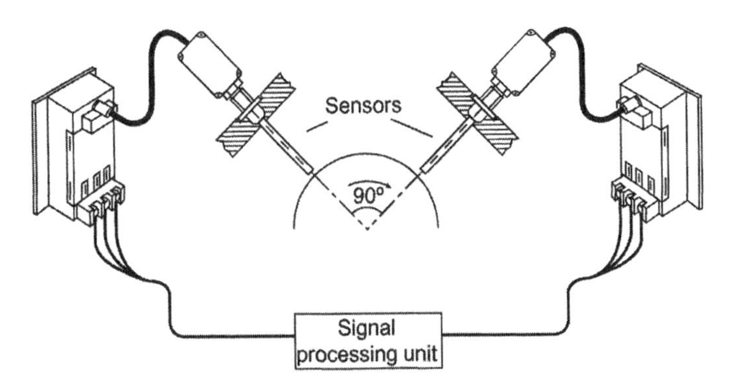

Fig. 9.7 Steam turbine vibration monitoring

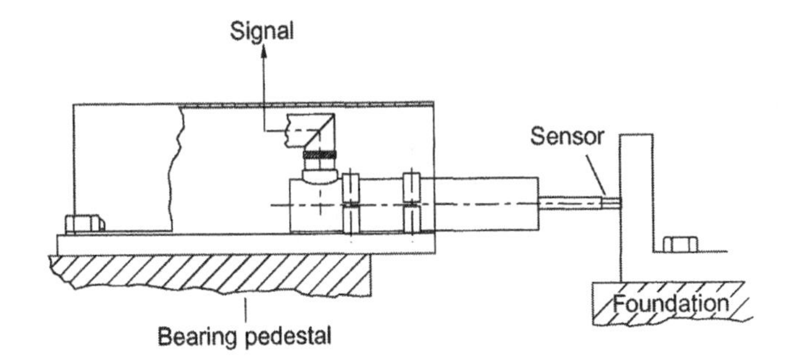

(a) Absolute expansion measurement

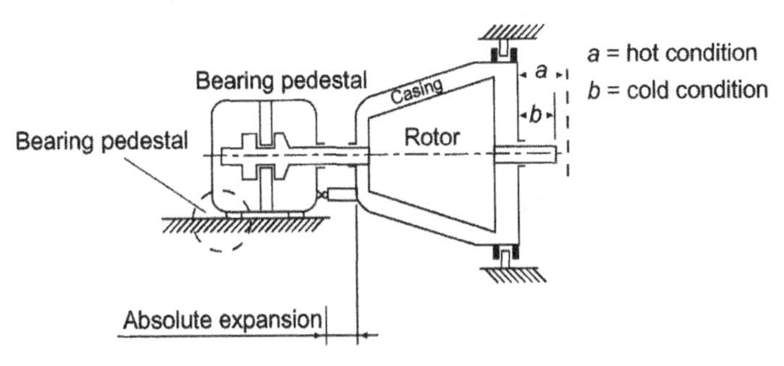

(b) Differential expansion measurement

Fig. 9.8 Steam turbine expansion measurement

Specifications and standards

Technical standards relating to steam turbines fall into three main categories.

The generalized technology standards

These provide a broad coverage of design, manufacture, and testing. Two important ones are API 611: (1989) *General purpose steam turbines for refinery service* and API 612: (1987) *Special purpose steam turbines for refinery service*. ASME/ANSI PTC (Power Test Codes) No. 6 is a related document group that complements the API standards.

Performance test standards

These cover only the performance testing of turbines under steaming conditions. They are used for performance verification after commissioning on site, but not for works testing. They are BS 5968: (1980) (similar to IEC 46-2) and BS 752: (1974) (similar to IEC 46-1) *Test code for acceptance tests*.

Procurement standards

The predominant document is BS EN 60045-1: (1993) *Steam turbine procurement* – identical to IEC 45-1. It has been recently updated and encompasses many of the modern practices governing the way in which steam turbines are specified and purchased. Some important parts of the content are:

- it provides clear guidance on governor characteristics and overspeed levels;
- vibration is addressed in two ways: bearing housing vibration using VDI 2056/ISO 2372: (1984) (using mm/s as the guiding parameter), and shaft vibration using ISO 7919: (1986) and the concept of relative displacement measurement;
- definitive requirements are stated for hydrostatic tests on the pressurized components of the turbine.

Turbine hydrostatic test

The predominant design criterion for turbine casings is the ability to resist hoop stress at the maximum operating temperature. For practical reasons, a hydrostatic test is carried out at ambient temperature. Some important points are given below.

- The test pressure is normally 150 per cent of the maximum allowable pressure the casing will experience in service. It is sometimes necessary to apply a multiplying factor to compensate for the difference in tensile strength of the steel between ambient and operating temperature. Practically, codes such as ASME Section VIII Division 1 are used to determine material stresses and the corresponding test pressure.

- Some types of casing (typically those that have been designed to very 'tight' stress criteria) are tested by sub-dividing the casing with steel diaphragms held in place by jacks. This enables the various regions of the casing to be tested at individual pressures that are more representative of the pressure gradient the casing experiences in use. Figure 9.9 shows such an arrangement.
- Hydrostatic pressure is maintained for a *minimum* of 30 min with two gauges fitted to identify any pressure drops.

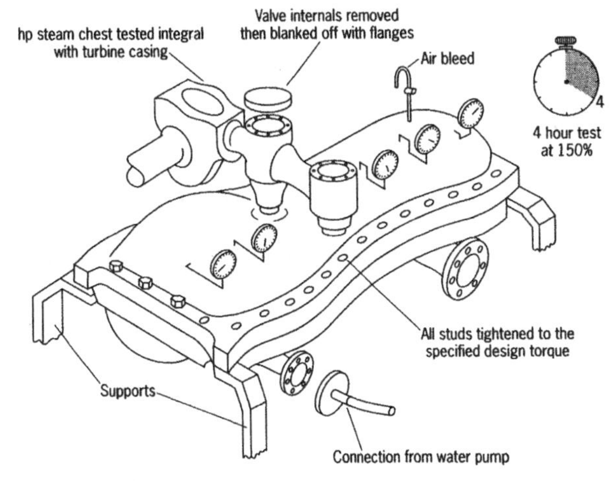

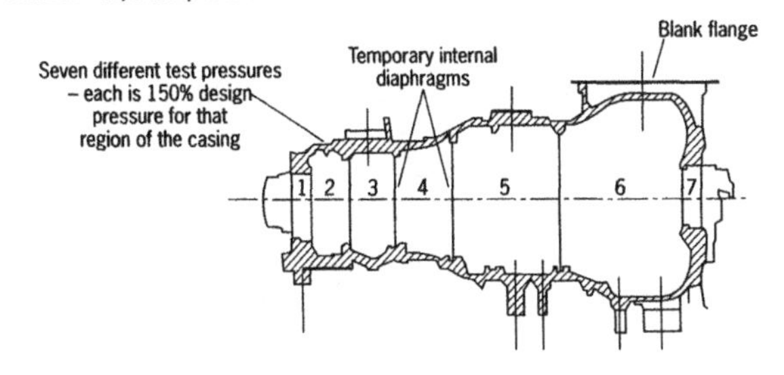

Fig. 9.9 Steam turbine casing hydrostatic test

Some important visual inspection points are:

- *Flange faces*. After the hydrostatic test, it is important to check the flatness of the flange-faces (using marking blue) to make sure no distortion has occurred. Pay particular attention to the inside edges; this is where distortion often shows itself first. Any 'lack of flatness' means that the faces must be skim milled.
- *Bolt-holes*. Visually check around all the flange bolt-holes for cracks.
- *Internal radii*. Check that small radii inside the casing have been well dressed and blended to minimize stress concentrations.
- *General surface finish*. There should be a good 'as-cast' finish on the inside of the casing without significant surface indentations. The visual inspection standard MSS-SP-55 is used as a broad guide.

Rotor tests

Steam turbine rotors are subject to dynamic balancing, overspeed, and tests on vibration assembly using similar techniques to those for gas turbine and gearbox rotors.

Dynamic balancing

This is carried out after the blades have been assembled, normally at low speed (400–500 r/min). Smaller HP and IP rotors will have two correction planes for adjustment weights, while large LP rotors have three.

API 611/612 specifies a maximum residual unbalance U per plane of

$$U\,(\text{g.mm}) = \frac{6350\ W(\text{kg})}{N(\text{r/min})}$$

where

W = journal load in kg
N = maximum continuous speed in r/min

ISO 1940 specifies its balance quality grade G2.5 for steam turbine rotors. A similar approach is adopted by VDI 2060.

Vibration

API 611/612 specifies vibration as an amplitude. The maximum peak-to-peak amplitude A (microns) is given by: $A\ (\mu\text{m}) = 25.4\ \sqrt{(12\,000/N)}$ with an absolute limit of 50 μm. BS EN 60045-1 adopts the same approach as other European turbine standards. Bearing housing vibration follows ISO 2372 (similar to VDI 2056) using a velocity V (r.m.s.) criterion of 2.8 mm/s. Shaft vibration is defined in relation to ISO 7919-1, which is a more complex approach.

Overspeed

Steam turbine rotor overspeed tests are carried out in a vacuum chamber to minimize problems due to windage. API 611/612 infers that a steam turbine rotor should be overspeed tested at 110 per cent of rated speed. BS EN 60045-1 places a maximum limit of 120 per cent of rated speed for the overspeed test. In practice, this is more usually 110 per cent.

Assembly tests

Most steam turbine clearances are measured before fitting the outer turbine casing top half. Figure 9.10 shows the locations at which the main clearances are taken and gives indicative values for a double-casing type HP turbine. Note the following points.

- *Gland clearances* Radial and axial clearances are normally larger at the low-pressure (condenser) end. The readings should be confirmed at four diametral positions.
- *Nozzle casing and balance piston seals* The axial clearances are generally approximately three times the radial clearances.
- *Blade clearances* These are measured using long (300–400 mm) feeler gauges to take clearance measurements at the less accessible radial locations. Note how the radial and axial clearances (and the allowable tolerances) increase towards the low-pressure end.

Radial clearances for the rotating blades tend to be broadly similar to those for the stationary blades. However, lower temperature turbines in which the fixed blades are carried in cast steel diaphragms may have smaller clearances for the labyrinth seal between the diaphragm and the rotor (this is due to the high-pressure drop across the impulse stages).

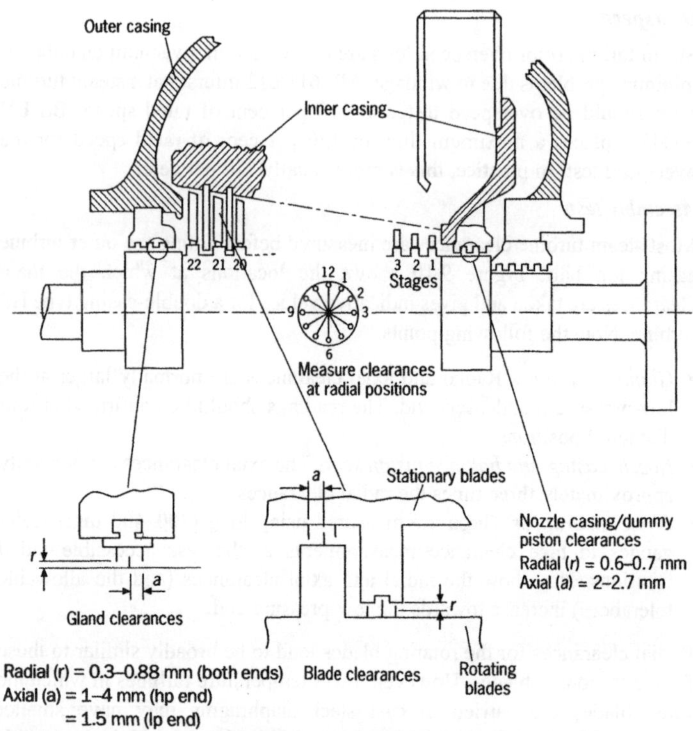

Stage	Rotating/stationary blades	
	Radial (r)	Axial (a)
1–4	0.9 ± 0.3 mm	4 ± 1.5 mm
5–9	0.9 ± 0.3 mm	5 ± 1.6 mm
10–14	1.0 ± 0.35 mm	5 ± 1.6 mm
15–18	1.2 ± 0.35 mm	6 ± 1.6 mm
19–22	1.4 ± 0.40 mm	6 ± 1.6 mm

Fig. 9.10 Steam turbine – typical HP turbine clearances

Useful standards

Table 9.1 contains published technical standards with particular reference to turbines.

Table 9.1 Technical standards – turbines

Standard	*Title*	*Status*
BS 3135: 1989, ISO 2314: 1989	Specification for gas turbine acceptance test.	Current
BS 3863: 1992, ISO 3977: 1991	Guide for gas turbines procurement.	Current, work in hand
BS 5671: 1979, IEC 60545: 1976	Guide for commissioning, operation, and maintenance of hydraulic turbines.	Current
BS 5860: 1980, IEC 60607: 1978	Method for measuring the efficiency of hydraulic turbines, storage pumps, and pump turbines (thermodynamic method).	Current
BS 7721: 1994, ISO 10494: 1993	Gas turbines and gas turbine sets. Measurement of emitted airborne noise. Engineering/survey method.	Current
BS 7854-2: 1996, ISO 10816-2: 1996	Mechanical vibration. Evaluation of machine vibration by measurements on non-rotating parts. Large land-based steam turbine generator sets in excess of 50 MW.	Current, work in hand
BS 7854-3: 1998, ISO 10816-3: 1998	Mechanical vibration. Evaluation of machine vibration by measurements on non-rotating parts. Industrial machines with nominal power above 15 kW and nominal speeds between 120 r/min and 15 000 r/min when measured in situ.	Current
BS 7854-4: 1998, ISO 10816-4: 1998	Mechanical vibration. Evaluation of machine vibration by measurements on non-rotating parts. Gas turbine driven sets excluding aircraft derivatives.	Current
BS ISO 7919-2: 1996	Mechanical vibration of non-reciprocating machines. Measurements on rotating shafts and evaluation criteria. Large land-based steam turbine generator sets.	Current, work in hand

Table 9.1 Cont.

BS ISO 7919-3: 1996	Mechanical vibration of non-reciprocating machines. Measurements on rotating shafts and evaluation criteria. Coupled industrial machines.	Current
BS ISO 7919-4: 1996	Mechanical vibration of non-reciprocating machines. Measurements on rotating shafts and evaluation criteria. Gas turbine sets.	Current
BS ISO 7919-5: 1997	Mechanical vibration of non-reciprocating machines. Measurements on rotating shafts and evaluation criteria. Machine sets in hydraulic power generating and pumping plants.	Current
BS ISO 11042-1: 1996	Gas turbines. Exhaust gas emission. Measurement and evaluation.	Current
BS ISO 11042-2: 1996	Gas turbines. Exhaust gas emission. Automated emission monitoring.	Current
BS ISO 11086: 1996	Gas turbines. Vocabulary.	Current
BS ISO 14661: 2000	Thermal turbines for industrial applications (steam turbines, gas expansion turbines). General requirements.	Current
BS IEC 61366-1: 1998	Hydraulic turbines, storage pumps, and pump turbines. Tendering documents. General and annexes.	Current
BS IEC 61366-2: 1998	Hydraulic turbines, storage pumps, and pump turbines. Tendering documents. Guidelines for technical specifications for Francis turbines.	Current
BS IEC 61366-3: 1998	Hydraulic turbines, storage pumps, and pump turbines. Tendering documents. Guidelines for technical specifications for Pelton turbines.	Current
BS IEC 61366-4: 1998	Hydraulic turbines, storage pumps, and pump turbines. Tendering documents. Guidelines for technical specifications for Kaplan and propeller turbines.	Current
BS IEC 61366-5: 1998	Hydraulic turbines, storage pumps, and pump turbines. Tendering documents. Guidelines for technical specifications for tubular turbines.	Current

Table 9.1 Cont.

BS IEC 61366-6: 1998	Hydraulic turbines, storage pumps, and pump turbines. Tendering documents. Guidelines for technical specifications for pump turbines.	Current
BS IEC 61366-7: 1998	Hydraulic turbines, storage pumps, and pump turbines. Tendering documents. Guidelines for technical specifications for storage pumps.	Current
BS EN 45510-2-6: 2000	Guide for the procurement of power station equipment. Electrical equipment. Generators.	Current
BS EN 45510-5-1: 1998	Guide for the procurement of power station equipment. Steam turbines.	Current
BS EN 45510-5-2: 1998	Guide for the procurement of power station equipment. Gas turbines.	Current
BS EN 45510-5-3: 1998	Guide for the procurement of power station equipment. Wind turbines.	Current
BS EN 45510-5-4: 1998	Guide for the procurement of power station equipment. Hydraulic turbines, storage pumps, and pump turbines.	Current
BS EN 45510-6-4: 2000	Guide for the procurement of power station equipment. Turbine auxiliaries. Pumps.	Current
BS EN 45510-6-9: 2000	Guide for the procurement of power station equipment. Turbine auxiliaries. Cooling water systems.	Current
BS EN 60034-3: 1996	Rotating electrical machines. Specific requirements for turbine-type synchronous machines.	Current
BS EN 60041: 1995	Field acceptance tests to determine the hydraulic performance of hydraulic turbines, storage pumps, and pump turbines.	Current
BS EN 60045-1: 1993, IEC 60045-1: 1991	Guide to steam turbine procurement.	Current
BS EN 60953-1: 1996, IEC 60953-1: 1990	Rules for steam turbine thermal acceptance tests. High accuracy for large condensing steam turbines.	Current
BS EN 60953-2: 1996, IEC 60953-2: 1990	Rules for steam turbine thermal acceptance tests. Wide range of accuracy for various types and sizes of turbines.	Current

Table 9.1 Cont.

BS EN 60994: 1993, IEC 60994: 1991	Guide for field measurement of vibrations and pulsations in hydraulic machines (turbines, storage pumps, and pump turbines).	Current
BS EN 60995: 1995, IEC 60995: 1991	Determination of the prototype performance from model acceptance tests of hydraulic machines with the consideration of scale effects.	Current
DD ENV 61400-1: 1995	Wind turbine generator systems. Safety requirements.	Current
BS EN 61400-2: 1996, IEC 61400-2: 1996	Wind turbine generator systems. Safety of small wind turbines.	Current
BS EN 61400-11: 1999, IEC 61400-11: 1998	Wind turbine generator systems. Acoustic noise measurement techniques.	Current, work in hand
BS EN 61400-12: 1998, IEC 61400-12: 1998	Wind turbine generator systems. Wind turbine power performance testing.	Current
95/701797 DC	Technical report for the nomenclature of hydraulic machinery (IEC/CD4/112/CDV).	Current, draft for public comment
95/713333 DC	Gas turbines. Procurement. Part 1. General and definitions (ISO/DIS 3977-1).	Current, draft for public comment
95/713334 DC	Gas turbines. Procurement. Part 2. Standard reference conditions and ratings (ISO/DIS 3977-2).	Current, draft for public comment
96/704522 DC	Gas turbines. Procurement. Part 11. Reliability, availability, maintainability, and safety (ISO/DIS 3977-11).	Current, draft for public comment
97/703272 DC	Hydraulic turbines, storage pumps, and pump turbines. Hydraulic performance. Model acceptance tests (IEC 193-2).	Current, draft for public comment
97/704872 DC	Gas turbines. Procurement. Part 7. Technical information (ISO/CD 3977-7).	Current, draft for public comment
97/704873 DC	Gas turbines. Procurement. Part 8. Inspection, testing, installation, and commissioning (ISO/CD 3877-8).	Current, draft for public comment
97/710296 DC	Gas turbines. Procurement. Part 6. Combined cycles (ISO 3977-6).	Current, draft for public comment

Table 9.1 Cont.

98/709429 DC	ISO/CD 3977-4.2. Gas turbines. Procurement. Part 4. Fuels and procurement (ISO/CD 3977-4.2).	Current, draft for public comment
98/711877 DC	Centrifugal pumps for petroleum, heavy-duty chemical, and gas industries services (ISO/DIS 13709).	Current, draft for public comment
98/716323 DC	Mechanical vibration. Evaluation of machine vibration by measurements on non-rotating parts. Part 5. Machine sets in hydraulic power generating and pumping plants (ISO/DIS 10816-5).	Current, draft for public comment
99/200755 DC	IEC 61400-22. Wind turbine certification (IEC Document 88/102/CD).	Current, draft for public comment
99/204720 DC	IEC 61400-23 TS Ed. 1. Wind turbine generator systems. Part 23. Full-scale structural testing of rotor blades for WTGSs (IEC Document 88/116/CDV).	Current, draft for public comment
99/710366 DC	IEC 4/155/CD. Hydraulic turbines. Testing of control systems.	Current, draft for public comment
00/200785 DC	IEC 61400-13. Wind turbine generator systems. Part 13. Measurement of mechanical loads (IEC Document 88/120/CDV).	Current, draft for public comment
00/702245 DC	ISO 10437. Petroleum and natural gas industries. Special purpose steam turbines for refinery service.	Current, draft for public comment
00/704175 DC	ISO/DIS 7919-2. Mechanical vibration. Evaluation of machine vibration by measurements on rotating shafts. Part 2. Land-based steam turbines and generators in excess of 50 MW with normal operating speeds of 1500 r/min, 1800 r/min, 3000 r/min, and 3600 r/min.	Current, draft for public comment

Table 9.1 Cont.

00/704176 DC	ISO/DIS 10816-2. Mechanical vibration. Evaluation of machine vibration by measurements on non-rotating parts. Part 2. Large land-based steam turbines and generators in excess of 50 MW with normal operating speeds of 1500 r/min, 1800 r/min, 3000 r/min, and 3600 r/min.	Current, draft for public comment
00/704290 DC	IEC 60953-3/Ed. 1 Rules for steam turbine thermal acceptance tests. Part 3. Thermal performance verification tests of retrofitted steam turbines.	Current, draft for public comment
00/704918 DC	ISO/DIS 3977-6. Gas turbines. Procurement. Part 6. Combined cycles.	Current, draft for public comment
00/712064 DC	ISO/DIS 102: 2000. Aircraft. Gravity filling orifices and nozzles.	Current, draft for public comment
BS 132: 1983	Guide for steam turbines procurement.	Withdrawn, superseded
BS 489: 1983	Specification for turbine oils.	Withdrawn, revised
BS 752: 1974	Test code for acceptance of steam turbines.	Withdrawn, superseded
BS 3135: 1975, ISO 2314-1973	Specification for gas turbines: acceptance tests.	Withdrawn, revised
BS 3853: 1966	Specification for mechanical balancing of marine main turbine machinery.	Withdrawn, superseded
BS 3863: 1979, ISO 3977-1978	Guide for gas turbines procurement.	Withdrawn, revised
BS 5000: Part 2: 1988	Rotating electrical machines of particular types or for particular applications. Specification for turbine-type synchronous machines.	Withdrawn, revised
BS 5968: 1980	Methods of acceptance testing of industrial-type steam turbines.	Withdrawn, superseded

9.2 Gas turbines – aeroderivatives

Although there are many variants of gas turbine-based aeroderivative engines, they operate using similar principles. Air is compressed by an axial flow or centrifugal compressor. The highly compressed air then passes to a combustion chamber where it is mixed with fuel and ignited. The mixture of air and combustion products expands into the turbine stage, which in turn provides the power through a coupling shaft to drive the compressor. The expanding gases then pass out through the engine tailpipe, providing thrust, or can be passed through a further turbine stage to drive a propeller or helicopter rotor. For aeronautical applications the two most important criteria in engine choice are thrust (or power) and specific fuel consumption. Figure 9.11 shows an outline of the main types and Table 9.2 gives the terminology.

Table 9.2 Gas turbine propulsion terminology

Gas turbine (GT)	Engine comprising a compressor and turbine. It produces jet thrust and/or shaft 'horsepower' output via a power turbine stage.
Turbojet	A GT which produces only jet thrust (i.e. no power turbine stage). Used for jet aircraft.
Turboprop	A GT that produces shaft output and some jet thrust. Used for propeller-driven aircraft.
Afterburner	A burner which adds fuel to the later stages of a GT to give increased thrust. Used for military aircraft.
Pulsejet	A turbojet engine with an intermittent 'pulsed' thrust output.
Ramjet	An advanced type of aircraft GT which compresses the air using the forward motion (dynamic head) of the engine.
Rocket motor	A 'jet' engine that carries its own fuel and oxygen supply. Produces pure thrust when there is no available oxygen (e.g. space travel).

The simple turbojet

The simple turbojet derives all of its thrust from the exit velocity of the exhaust gas. It has no separate propeller or 'power' turbine stage. Performance parameters are outlined in Fig. 9.12. Turbojets have poor fuel economy and high exhaust noise. The fact that all the air passes through the engine core (i.e. there is no bypass) is responsible for the low propulsive efficiency, except at very high aircraft speed. The Concorde supersonic

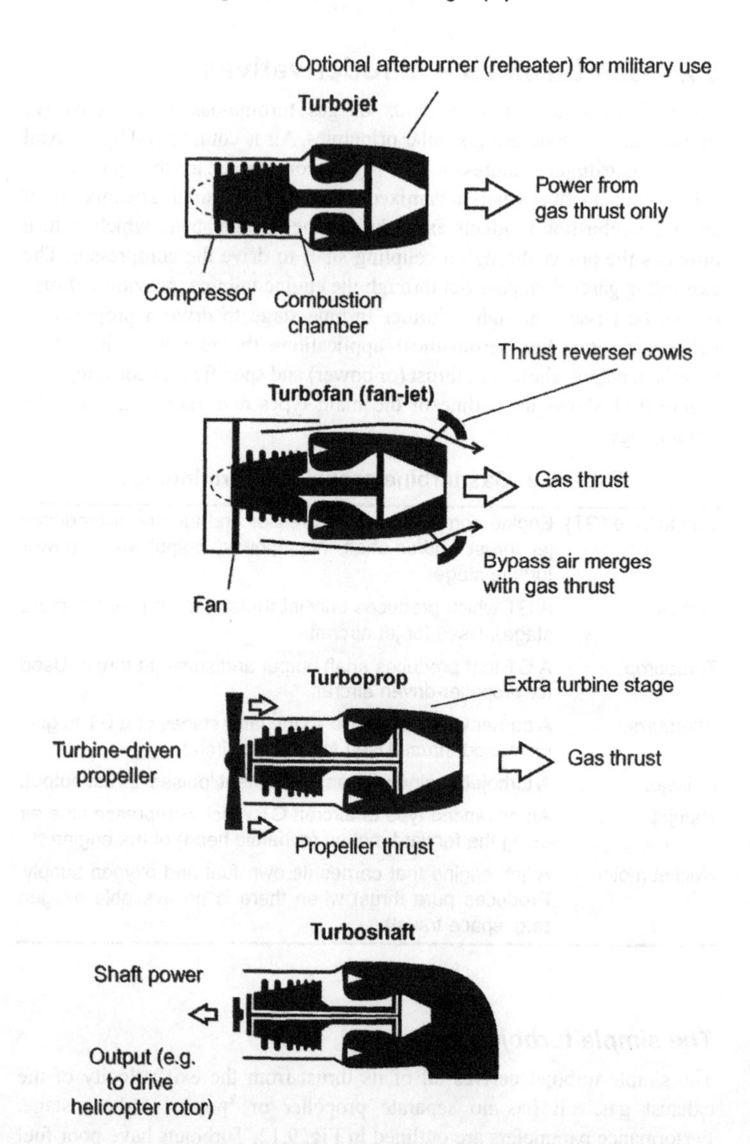

Fig. 9.11 Aero gas turbines – main types

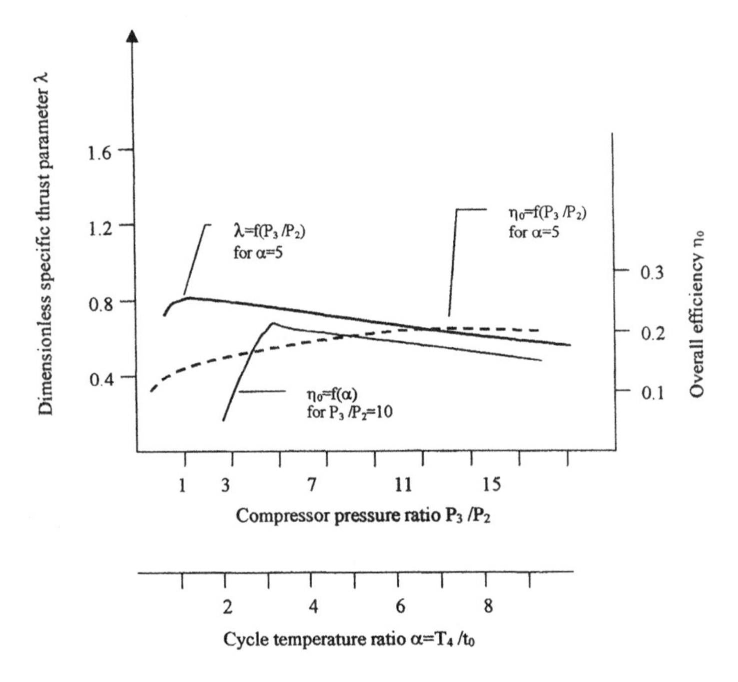

Fig. 9.12 Aero turbojet – typical performance parameters

transport (SST) aircraft is virtually the only commercial airliner that still uses the turbojet. By making the convenient assumption of neglecting Reynolds number, the variables governing the performance of a simple turbojet can be grouped as shown in Table 9.3.

Table 9.3 Turbojet performance parameter groupings

Non-dimensional group	*Uncorrected*	*Corrected*
Flight speed	$V_0/\sqrt{t_0}$	$V_0/\sqrt{\theta}$
RPM	$N/\sqrt{T}$	$N/\sqrt{\theta}$
Air flow rate	$\dot{W}_a/\sqrt{(T/D^2P)}$	$\dot{W}_a/\sqrt{(\theta/\delta)}$
Thrust	F/D^2P	F/δ
Fuel flow rate	$\dot{W}_f J\Delta H_c/D^2P\sqrt{T}$	$\dot{W}_f/\delta\sqrt{\theta}$

where

$\theta = T/T_{std} = T/519\ (T/288)$ = corrected temperature

$\delta = P/p_{std} = P/14.7\ (P/1.013 \times 10^5)$ = corrected pressure

$\dot{W}_f$ = fuel flow

Turbofan

Most large airliners and subsonic aircraft are powered by turbofan engines. Typical commercial engine thrust ratings range from 7000 lb (31 kN) to 90 000 lb (400 kN+), suitable for large aircraft such as the Boeing 747. The turbofan is characterized by an oversized fan compressor stage at the front of the engine which bypasses most of the air around the outside of the engine where it re-joins the exhaust gases at the back, increasing significantly the available thrust. A typical bypass ratio is 5–6 to 1. Turbofans have better efficiency than simple turbojets because it is more efficient to accelerate a large mass of air moderately through the fan to develop thrust, than to highly accelerate a smaller mass of air through the core of the engine to develop the same thrust. Figure 9.13 shows the basic turbofan and Fig. 9.14 its two- and three-spool variants. The two-spool arrangement is the most common, with a single-stage fan plus turbine on the low-pressure rotor and an axial compressor plus turbine on the high-pressure rotor. Many turbines are fitted with thrust-reversing cowls that act to reverse the direction of the slipstream of the fan bypass air.

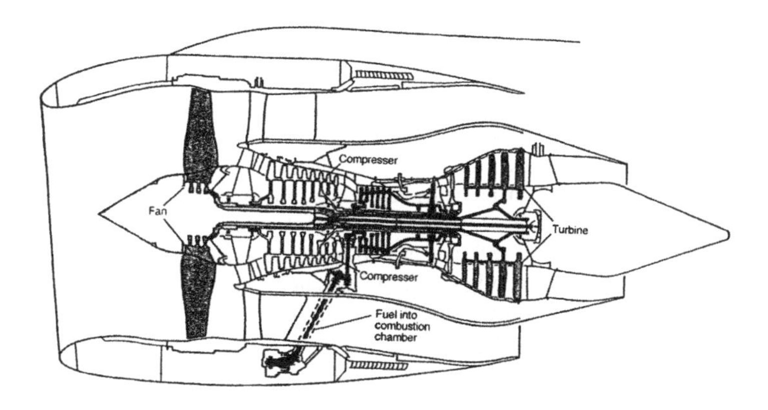

Fig. 9.13 The basic aero turbofan

Two-spool (most common aero engine configuration)

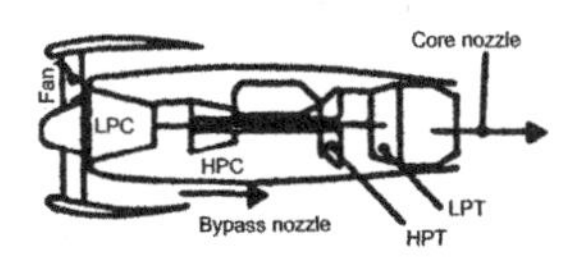

High-pressure spool – The hp turbine (HPT) drives the high-pressure compressor (HPC)

Low-pressure spool – The lp turbine (LPT) drives the low-pressure compressor (LPC)

Three-spool engine (Rolls Royce RB211)

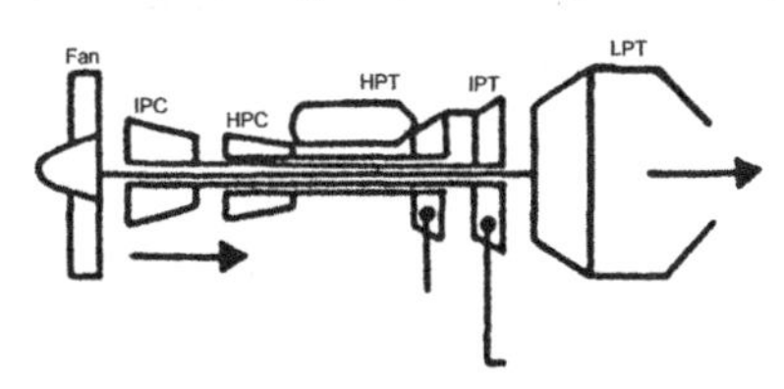

Fig. 9.14 Aero turbanfan – two- and three-spool variants

Turboprop

The turboprop configuration is typically used for smaller aircraft. The engine (see Fig. 9.11) uses a separate power turbine stage to provide torque to a forward-mounted propeller. The propeller thrust is augmented by gas thrust from the exhaust. Although often overshadowed by the turbofan, recent developments in propeller technology mean that smaller airliners such as the SAAB 2000 (2 × 4152 hp [3096 kW] turboprops) can compete on speed and fuel cost with comparably-sized turbofan aircraft. The most common turboprop configuration is a single shaft with centrifugal compressor and integral gearbox. Commuter airliners often use a two- or three-shaft 'free turbine' layout.

Propfans

Propfans are a modern engine arrangement specifically designed to achieve low fuel consumption. They are sometimes referred to as 'inducted' fan engines. The most common arrangement is a two-spool gas generator and aft-located gearbox driving a 'pusher' fan. Historically, low fuel prices have reduced the drive to develop propfans as commercially viable mainstream engines. Some Russian aircraft, such as the Anotov An-70 transport design, have been designed with propfans.

Turboshafts

Turboshaft engines are used predominantly for helicopters. A typical example, such as the Rolls-Royce Turbomeca RTM 32201, has a three-stage axial compressor directly coupled to a two-stage compressor turbine, and a two-stage power turbine. Drive is taken off the power turbine shaft, through a gearbox, to drive the main and tail rotor blades. Figure 9.11 shows the principle.

Ramjet

This is the crudest form of jet engine. Instead of using a compressor it uses the 'ram effect' obtained from its forward velocity to accelerate and pressurize the air before combustion. Hence, the ramjet must be accelerated to speed by another form of engine before it will start to work. Ramjet-propelled missiles, for example, are released from moving aircraft or accelerated to speed by booster rockets. A supersonic version is the 'scramjet' which operates on liquid hydrogen fuel.

Pulsejet

A pulsejet is a ramjet with an air inlet that is provided with a set of shutters fixed to remain in the closed position. After the pulsejet engine is launched, ram air pressure forces the shutters to open, and fuel is injected into the combustion chamber and burned. As soon as the pressure in the combustion chamber equals the ram air pressure, the shutters close. The gases produced by combustion are forced out of the jet nozzle by the pressure that has built up within the combustion chamber. When the pressure in the combustion chamber falls off, the shutters open again, admitting more air, and the cycle repeats.

Aero engine data

Table 9.4 shows indicative design data for commercially available aero engines from various manufacturers.

Table 9.4 Commercial aero engines – data tables

Company	*Allied signal*	*CFE*	*CFMI*	*General Electric (GE)*					*IAE (PW, RR, MTU, JAE)*		
Engine type/model	*LF507*	*CFE738*	*CFM 56 5C2*	*CF34 3A,3B*	*CF6 80A2*	*CF6 80C2-B2*	*CF6 80E1A2*	*GE 90 85B*	*V2500 A1*	*V2522 A5*	*V2533 A5*
Aircraft	BA146-300 Avro RJ	Falcon 2000	A340	Canadair RJ	A310-200 B767-200	B767-200ER	A330 B777-200/300		A320 A319	MD90-10/30 A319	A321-200
In service date	1991	1992	1994	1996	1981	1986	1995		1989	1993	1994
Thrust (lb)	7000	5918	31200	9220	60000	52500	67500	90000	25000	22000	33000
Flat rating (°C)	23.0	30.0	30.0		33.3	32.0	30.0	30.0	30.0	30.0	30.0
Bypass ratio	5.60	5.30	6.40						5.40	5.00	4.60
Pressure ratio	13.80	23.00	31.50	21.00	27.30	27.10	32.40	39.30	29.40	24.9	33.40
Mass flow (lb/s)	256	240	1065		1435	1650	1926	3037	781	738	848
SFC (lb/hr/lb)	0.406	0.369	0.32	0.35	0.35	0.32	0.33		0.35	0.34	0.37
Climb											
Max thrust (lb)			7580			12650		18000	5620	5550	6225
Flat rating (°C)									ISA+10	ISA+10	ISA+10
Cruise											
Altitude (ft)		40000	35000		35000	35000		35000	35000	35000	35000
Mach number		0.80	0.80		0.80	0.80		0.83	0.80	0.80	0.80
Thrust (lb)		1310			11045	12000			5070	5185	5725
Thrust lapse rate					0.221	0.229			0.202	0.2	0.174
Flat rating (°C)									ISA+10	ISA+10	ISA+10
SFC (lb/hr/lb)	0.414	0.645	0.545		0.623	0.576	0.562	0.545	0.581	0.574	0.574

Table 9.4 Cont.

Dimensions											
Length (m)	1.620	2.514	2.616	2.616	3.980	4.267	4.343	5.181	3.200	3.204	3.204
Fan diameter (m)	1.272	1.219	1.945	1.245	2.490	2.694	2.794	3.404	1.681	1.681	1.681
Basic eng. weight (lb)	1385	1325	5700	1670	8496	9399	10726	16644	5210	5252.0	5230.0
Layout											
Number of shafts	2	2	2	2	2	2	2	2	2	2	2
Compressor	various	1+5LP+-1CF	1+4LP 9HP	1F+14cHP	1+3LP 14HP	1+4LP 14HP	1+4LP 14HP	1+3LP 10HP	1+4LP 10HP	1+4LP 10HP	1+4LP 10HP
Turbine	2HP 2LP	2HP 3LP	1HP 5LP	2HP 4LP	2HP 4LP	2HP 5LP	2HP 5LP	2HP 6LP	2HP 5LP	2HP 5LP	2HP 5LP

Table 9.4 Cont.

Company	**Pratt and Whitney**					**Rolls-Royce**				**ZMKB**
Engine type/model	*PW4052*	*PW4056*	*PW4152*	*PW4168*	*PW4084*	*TRENT 772*	*TRENT 892*	*TAY 611*	*RB-211524H*	*D-436T1*
Aircraft	B767-200 &200ER	B747-400 767-300ER	A310	A330	B777	A330	B777	F100.70 Gulfst V	B747-400 B767-300	Tu-334-1 An 72,74
In service date	1986	1987	1986	1993	1994	1995		1988	1989	1996
Thrust (lb)	52200	56750	52000	68000	84000	71100	91300	13850	60600	16865
Flat rating (°C)	33.3	33.3	42.2	30.0	30.0	30.0	30.0	30.0	30.0	30.0
Bypass ratio	4.85	4.85	4.85	5.10	6.41	4.89	5.74	3.04	4.30	4.95
Pressure ratio	27.50	29.70	27.50	32.00	34.20	36.84	42.70	15.80	33.00	25.20
Mass flow (lb/s)	1705	1705	1705	1934	2550	1978	2720	410	1605	
SFC (lb/hr/lb)	0.351	0.359	0.348					0.430	0.563	
Climb										
Max thrust (lb)						15386	18020	3400	12726	
Flat rating (°C)						ISA+10	ISA+10	ISA+5	ISA+10	
Cruise										
Altitude (ft)	35000	35000		35000	35000	35000	35000	35000	35000	36089
Mach number	0.80	0.80		0.80	0.83	0.82	0.83	0.80	0.85	0.75
Thrust (lb)						11500	13000	2550	11813	3307
Thrust lapse rate						0.162	0.142	0.184	0.195	0.196
Flat rating (°C)						ISA+10	ISA+10		ISA+10	
SFC (lb/hr/lb)						0.565	0.557	0.690	0.570	0.610

Table 9.4 Cont.

Dimensions										
Length (m)	3.879	3.879	3.879	4.143	4.869	3.912	4.369	2.590	3.175	
Fan diameter (m)	2.477	2.477	2.477	2.535	2.845	2.474	2.794	1.520	2.192	1.373
Basic eng. weight (lb)	9400	9400	9400	14350	13700	10550	13133	2951	9670	3197
Layout										
Number of shafts	2	2	2	2	2	3	3	2	3	3
Compressor	1+4LP 11HP	1+4LP 11HP	1+4LP 11HP	1+5LP 11HP	1+6LP 11HP	1LP 8IP 6HP	1LP 8IP 6HP	1+3LP 12HP	1LP 7IP 6HP	1+1L 6I 7HP
Turbine	2HP 4LP	2HP 4LP	2HP 4LP	2HP 5LP	2HP 7LP	1HP 1IP 4LP	1HP 1IP 5LP	2HP 3LP	1HP 1IP 3LP	1HP 1IP 3LP

9.3 Gas turbines – industrial

There are a wide variety of gas turbines (GTs) that have been adapted for industrial use for power generation and process use.

Basic principles

Figure 9.15 shows the schematic arrangement of an industrial-type GT and the corresponding graphical representation of a temperature/enthalpy-entropy (T/h–s) diagram for four main variants: advance sequential combustion, single combustion, standard design, and aeroderivative type. Efficiency increases with the area of the 'enveloping' process curve.

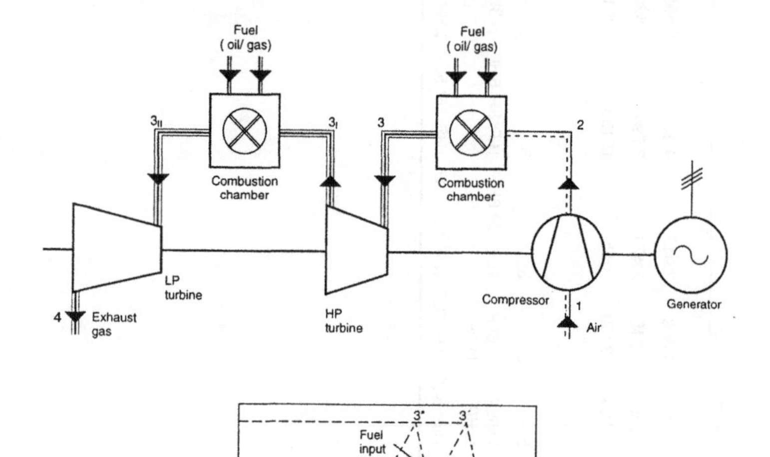

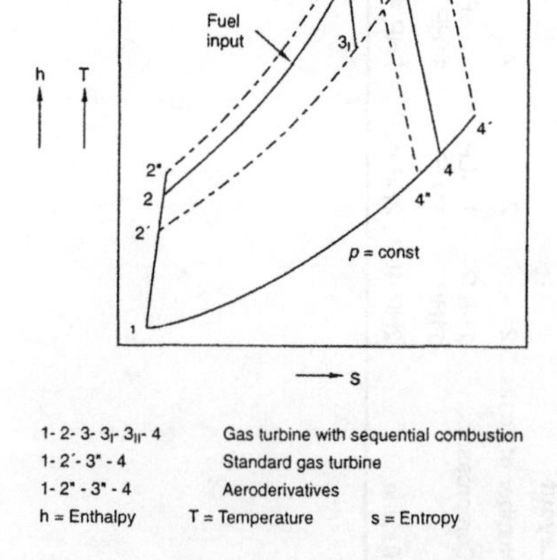

Fig. 9.15 Industrial gas turbine – schematic arrangement and T–s/h–s characteristics

Axial flow compressor characteristics

In many industrial GT designs, the combustion air is compressed by an axial flow compressor attached to the same shaft as the turbine stages. The blade stages increase the velocity of the air, then convert the resulting kinetic energy into 'pressure energy'. The power required to drive the compressor is derived from the power produced by the subsequent expansion of the gas, after combustion, through the turbine. Figure 9.16 shows the velocity relationships across a typical GT compressor stage.

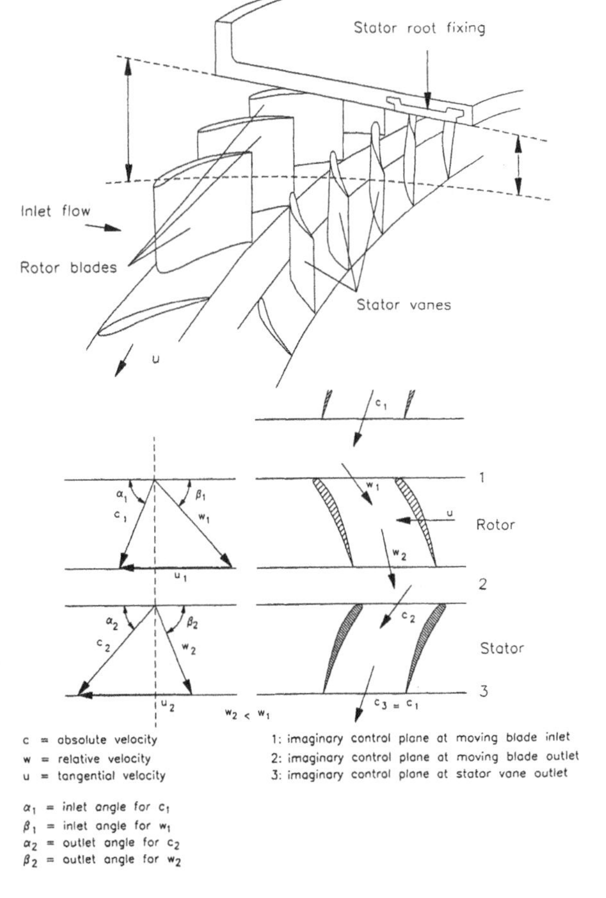

Fig. 9.16 Velocity relationships across a GT compressor stage

Axial flow turbine characteristics

Turbine blades are arranged in 'stages' that act like a convergent nozzle. Combustion gas enters the moving blade row with velocity c_1, which is resolved into relative and tangential velocity components w_1 and u_1, respectively. The effect of the blades is to increase the relative fluid velocity component (to w_2) without any change to the tangential velocity component u. Such velocity diagrams (Fig. 9.17) are used to develop the optimum geometry of blade profiles. Similar diagrams showing axial, radial, and tangential *force* components are used to design the overall ruling section requirements and strength of the rotating and stationary blades.

Gas turbines – major components

Land-based gas turbines for power generation, etc. comprise various major component systems (see Fig. 9.18). Many are dual fired, i.e. can burn either natural gas or light distillate oil.

Air intake system

The air intake system comprises an arrangement of mechanical shutters, filters, silencers, and safety flap valves (see Fig. 9.19). A compressed pulse air system is installed to provide periodic cleaning of the filters. Anti-icing hot air can be supplied from the GT compressor stages to prevent freezing of the inlet regions in cold weather.

The compressor stages

The compressor uses a combination of rotating and stationary blades to compress the cleaned inlet air prior to combustion. Each pair of stationary and rotating blades is termed a 'stage' and there are typically up to 24 stages in large, land-based GTs. Compressor blades are located in circumferential grooves, separated by spacers.

Variable guide vanes (VGVs)

VGVs are movable vanes, installed in rows, which regulate the volume of air that flows through the compressor.

Blow-off valves

Large blow-off valves are fitted to (normally) two stages of the compressor. These are necessary during low rotor speed, start-up and shut-down conditions in order to compensate for flow mismatch between the compressor and turbine stages by blowing off excess air.

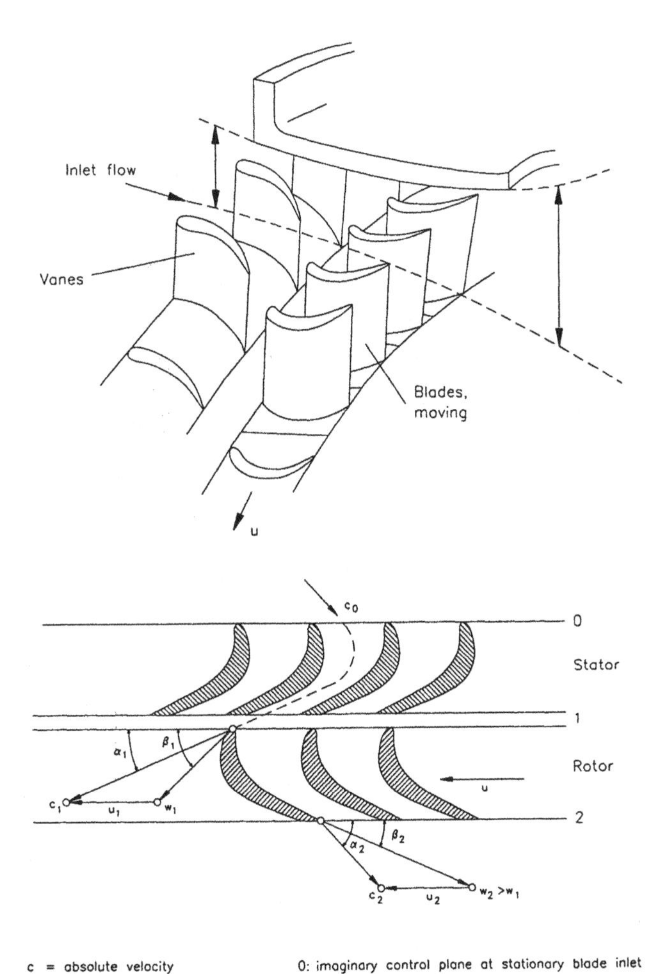

c = absolute velocity
w = relative velocity
u = tangential velocity

α_1 = inlet angle for c_1
β_1 = inlet angle for w_1
α_2 = outlet angle for c_2
β_2 = outlet angle for w_2

0: imaginary control plane at stationary blade inlet
1: imaginary control plane at moving blade inlet
2: imaginary control plane at moving blade outlet

Fig. 9.17 Velocity relationships across a GT turbine 'reaction' stage

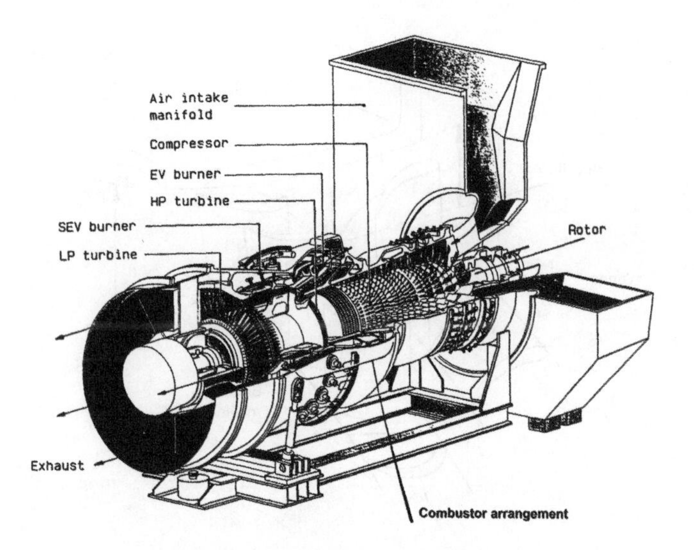

Fig. 9.18 Industrial GT – general view

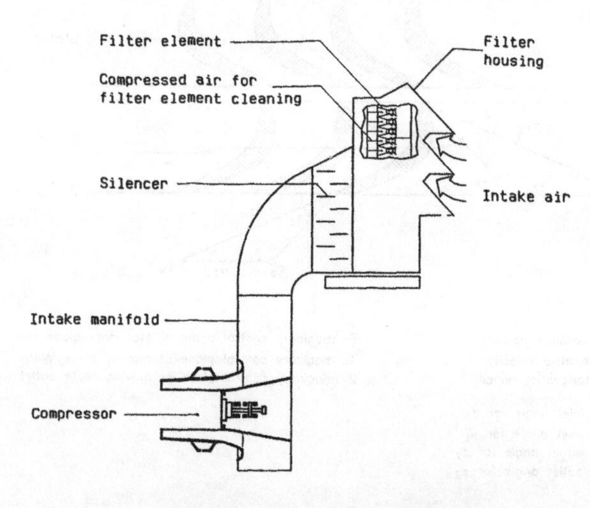

Fig. 9.19 GT air intake system

Diffuser

The diffuser is a ring-shaped assembly situated after the last compressor stage and before the combustion stages.

Combustion system

Advanced high-efficiency designs of land-based GTs use a two-stage sequential combustion system, loosely termed the environmental vane (EV) stage and a sequential environmental vane (SEV) stage (see Fig. 9.20). Both combustion chambers are cooled by air bled off the compressor stages and distributed through grooves arranged in an annular pattern around the GT casing. Combustion air flows into the EV combustion zone through inlet slots and mixes with the fuel gas, which enters via rows of fine holes at the edge of the slots (or the fuel oil which is sprayed in through a lance). The fuel is ignited by a separate ignition gas system and electric ignition torches. The SEV combustion stage is fed with hot gas from the EV stage. Additional fuel is admitted by the SEV burners which reheat the gas. Separate ignition is not normally required in the SEV stage as the gases are already hot enough to ignite the SEV stage fuel. A typical design of 200 MW+ power generation GT will have 28–32 EV burners and 22–28 SEV burners.

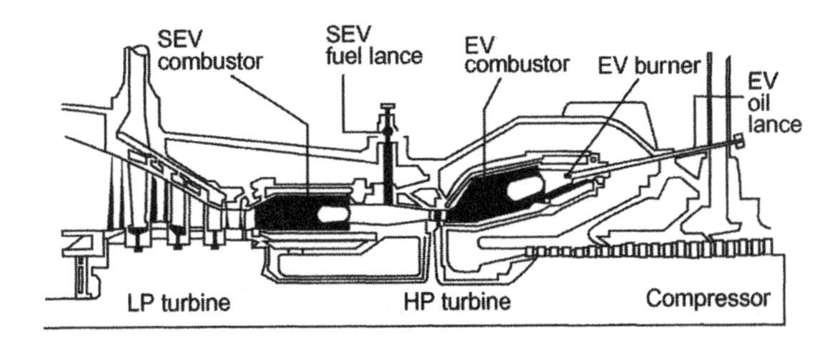

Fig. 9.20 GT sequential combustion arrangement

GT casing

GTs use horizontally split cast steel casings to enclose the vane carriers, which hold the stator blade assemblies, and the other stationary and rotating components. The casing incorporates a complex arrangement of cooling channels, structural reinforcement, and heat/noise insulation. The internal vane carriers are also split horizontally and protected by heat shields.

GT bearings

A series of bearings support the GT shaft (see Fig. 9.21). The turbine 'hot end' journal bearing supports the rotor in the radial direction, while the compressor 'cold end' bearing also takes axial thrust. Bearings are air-cooled using air bled off from various compressor stages. Sensors (see Fig. 9.22) detect both axial and radial rotor movement and vibration.

Exhaust gas system

The most common system for land-based GTs is for the hot combustion gas to be exhausted to a heat recovery steam generator (HRSG) – a large waste-heat boiler incorporated into a combined cycle system. Gas exhausts from the GT via an insulated diffuser and silencer, before entering the HRSG or stack (chimney).

Gas turbine inspections and testing

Acceptance guarantees

Acceptance guarantees for gas turbines are an uneasy hybrid of explicit and inferred requirements. Most contract specifications contain four explicit performance guarantee requirements: power output, net specific heat rate, auxiliary power consumption, and NO_x emission level. These are heavily qualified by a set of correction curves that relate to the various differences between reference conditions and those experienced at the installation site. The main ones are:

- governing characteristics;
- overspeed settings;
- vibration and critical speeds;
- noise levels.

Specifications and standards

The following standards are in common use in the GT industry.

- ISO 3977: (1991) *Guide for gas turbine procurement*, identical to BS 3863: 1992. This is a guidance document, useful for information on definitions of cycle parameters and for explaining different open and closed cycle arrangements.
- ISO 2314 *Gas turbine acceptance tests*, identical to BS 3135. This is not a step-by-step procedure for carrying out a no-load running test, but contains mainly technical information on parameter variations and measurement techniques for pressures, flows, powers, etc.
- ANSI/ASME PTC 22 *Gas turbine power plants*. This is one of the power test codes (PTC) family of standards. Its content is quite limited, covering broadly the same area as ISO 2314, but in less detail.

Journal and thrust bearing compressor end

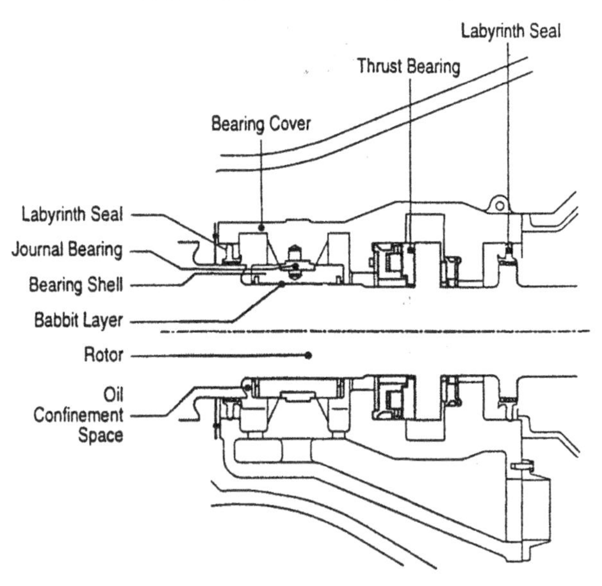

Journal bearing turbine end

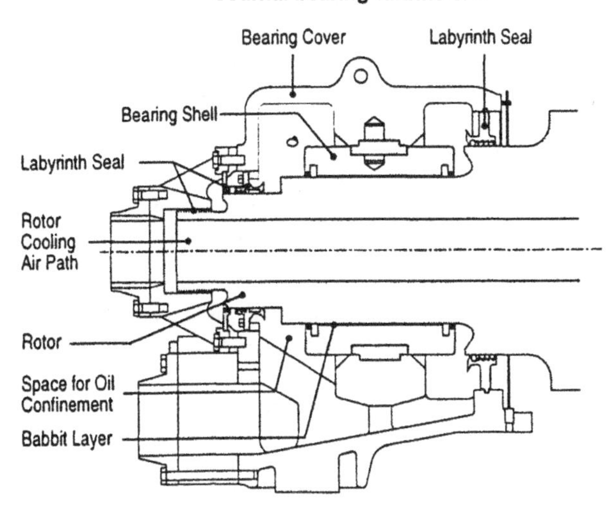

Fig. 9.21 GT shaft bearings

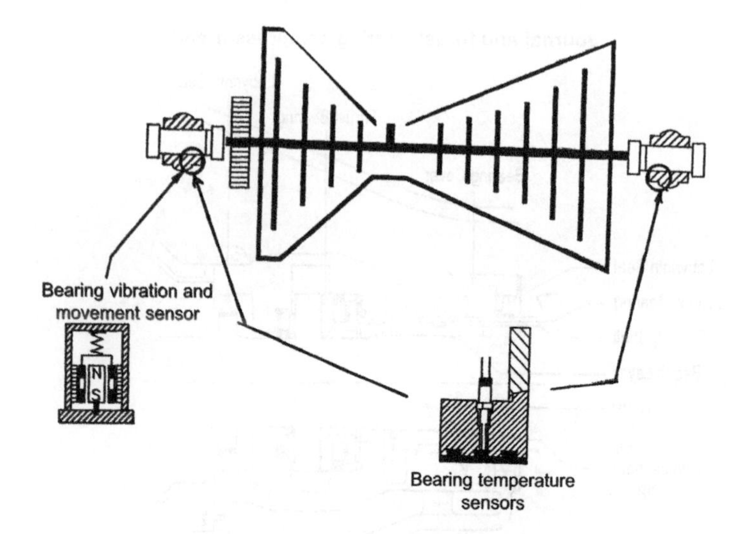

Fig. 9.22 GT bearing sensors

- API 616 *Gas turbines for refinery service*. In the mould of most API standards, this provides good technical coverage. There is a bit of everything to do with gas turbines.
- ISO 1940/1: (1986) *Balance quality requirements of rigid rotors. Part I – Method for determination of possible residual inbalance* (identical to BS 6861 Part 1 and VDI 2060) covers balancing of the rotor. It gives acceptable unbalance limits.
- ISO 10494: (1993) *Gas turbines and gas turbine sets; measurement of emitted airborne noise – engineering survey method* (similar to BS 7721) and ISO 1996 are standards relating to GT noise levels.

Vibration standards

- Bearing housing vibration is covered by VDI 2056 (group T). This is a commonly used standard for all rotating machines. It uses vibration velocity (mm/s) as the deciding parameter.
- Shaft vibrations using direct-mounted probes are covered by API 616 or ISO 7919/1 (also commonly used for other machines). The measured vibration parameter is amplitude. VDI 2059 (Part 4) is sometimes used, but it is a more theoretical document that considers the concept of non-sinusoidal vibrations.

Rotor runout measurement

The measured parameter is Total Indicated Runout (TIR). This is the biggest recorded difference in dial test indicator (d.t.i.) reading as the rotor is turned through a complete revolution. Figure 9.23 shows a typical results format.

- *Acceptance limits.* The maximum acceptable TIR is usually defined by the manufacturer rather than specified directly by a technical standard. The tightest limit is for the bearing journals (typically 10–15 μm). Radial surfaces of the turbine blade discs should have a limit of 40–50 μm. Axial faces of the discs often have a larger limit, perhaps 70–90 μm. The exact limits used depend on the design.

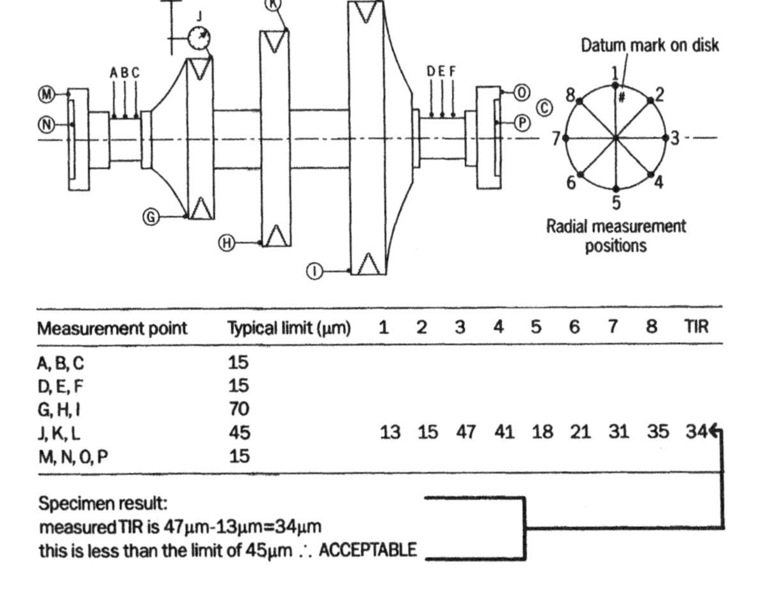

Measurement point	Typical limit (μm)	1	2	3	4	5	6	7	8	TIR
A, B, C	15									
D, E, F	15									
G, H, I	70									
J, K, L	45	13	15	47	41	18	21	31	35	34
M, N, O, P	15									

Specimen result:
measured TIR is 47μm-13μm=34μm
this is less than the limit of 45μm ∴ ACCEPTABLE

Fig. 9.23 GT rotor runout measurement

Rotor dynamic balancing

Gas turbine rotors are all subjected to dynamic balancing, normally with all the blades installed. Procedures can differ slightly; turbines that have separate compressor and turbine shafts may have these balanced separately (this is common on larger three-bearing designs) although some manufacturers prefer to balance the complete rotor assembly. The important parameter is the limit of acceptable unbalance expressed per correction plane (as in ISO 1940) in gramme metres (g.m). Figure 9.24 shows the arrangement.

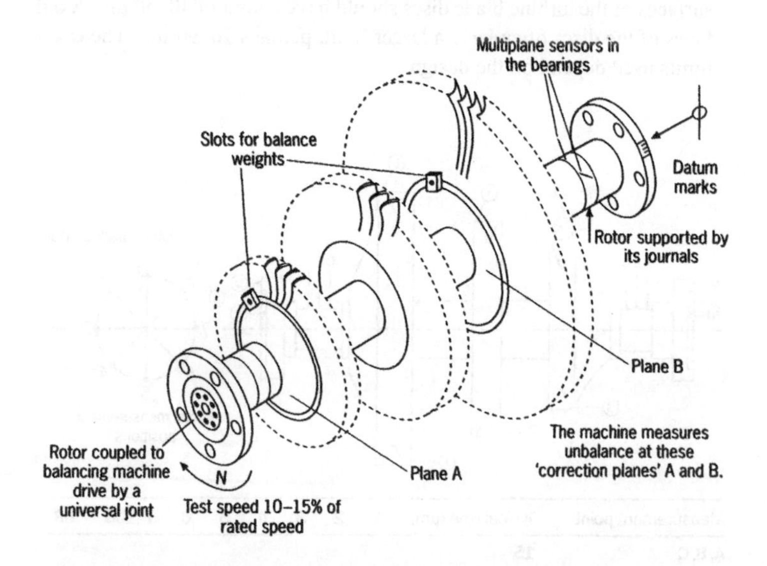

Fig. 9.24 Two-plane (dynamic) balancing of a GT rotor

Rotor overspeed test

Gas turbine rotors are subjected to an overspeed test with all the compressor and turbine blades in position. The purpose is to verify the mechanical integrity of the stressed components without stresses reaching the elastic limit of the material. It also acts as a check on vibration characteristics at the rated and overspeed condition. The test consists of running the rotor at 120 per cent rated speed for three minutes. Drive is by a large electric motor and the test is performed in a concrete vacuum chamber to eliminate windage. Full vibration monitoring to VDI 2056 or API 616 is performed, as mentioned earlier.

Blade clearance checks

The purpose is to verify running clearances between the ends of the rotor blades and the inside of the casing. Clearances that are too large will result in reduced stage efficiency. If the clearances are too tight, the blades may touch the inside of the casing and cause breakage, particularly at the compressor end. Figure 9.25 shows the arrangement. Indicative clearances (measured using slip gauges) are:

- compressor end stage 1–5 – 1.6 to 2.0 mm
- compressor end stage 9–16 – 1.8 to 2.4 mm
- compressor end stage 16+ – 2.0 to 2.4 mm
- turbine end – 4.0 to 4.5 mm
- rotor axial position (end clearance of last blades) – 7.0 to 8.0 mm

Figure 9.26 shows the profile of the GT no-load running test.

Noise measurement

Most contract specifications require that the GT be subject to a noise measurement check. The main technical standard relating to GT noise testing is ISO 10494: (1993) *Measurement of airborne noise* (equivalent to BS 7721). This is referred to by the GT procurement standard ISO 3977 and contains specific information about measuring GT noise levels. Noise measurement principles and techniques are common for many types of engineering equipment, so the following general technical explanations can be applied equally to diesel engines, gearboxes, or pumps.

Principles

It is easiest to think of noise as airborne pressure pulses set up by a vibrating surface source. It is measured by an instrument that detects these pressure changes in the air and then relates this measured sound pressure to an accepted 'zero' level. Because a machine produces a mixture of frequencies (termed 'broad-band' noise), there is no single noise measurement that will describe fully a noise emission. Two measurements are normally taken:

- *The 'overall noise' level* This is a colloquial term for what is properly described as the 'A-weighted sound pressure level'. It incorporates multiple frequencies and weights them according to a formula which results in the best approximation of the loudness of the noise. This is displayed as a single instrument reading expressed as decibels – in this case dB(A).

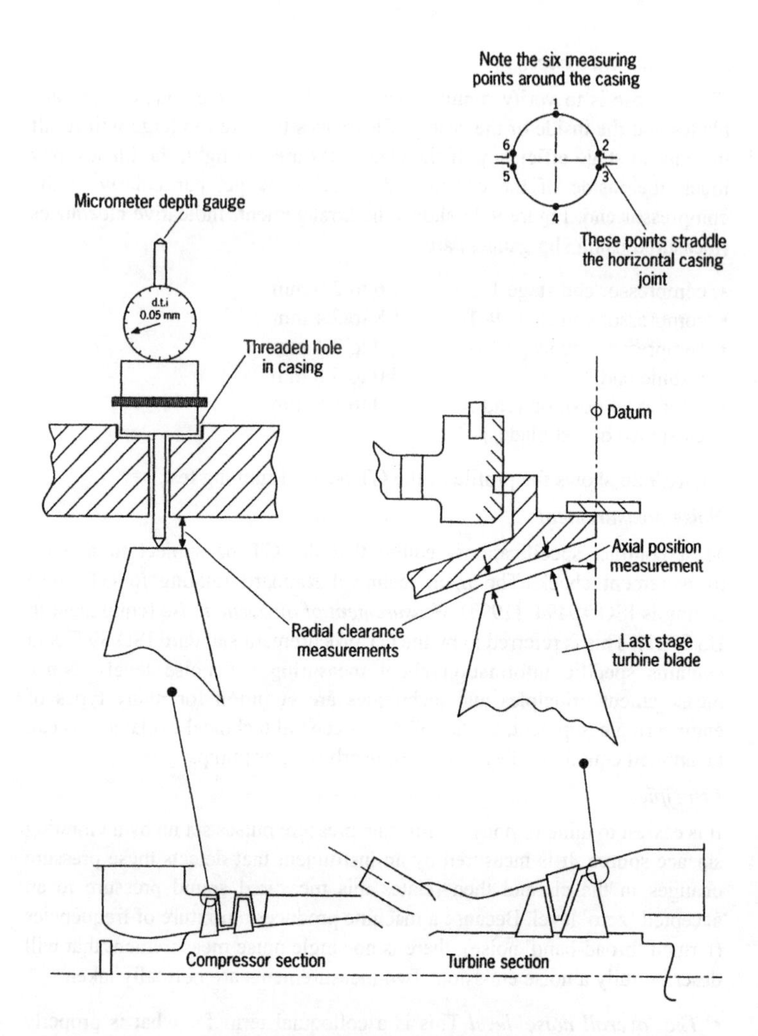

Fig. 9.25 GT clearance checks

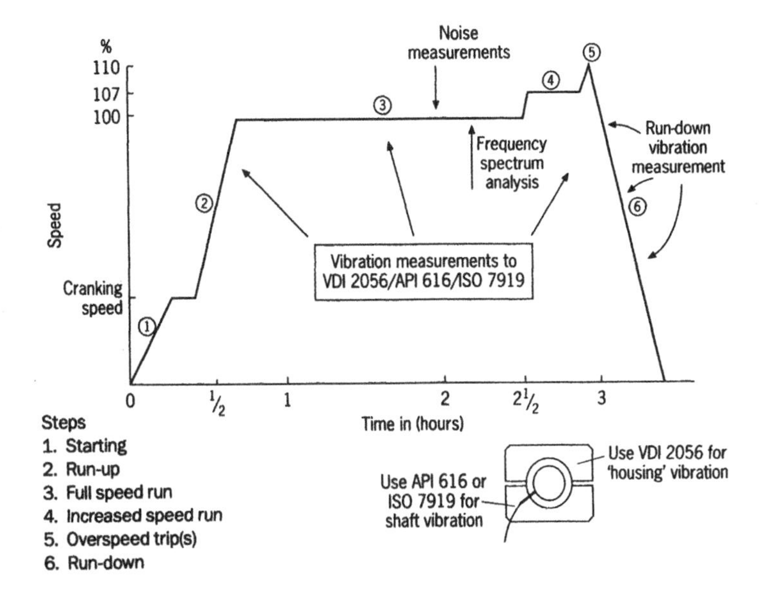

Fig. 9.26 GT no-load run test

- *'Frequency band' sound pressure level* This involves measuring the sound pressure level in a number of frequency bands. These are arranged in either octave or one-third octave bands in terms of their mid-band frequency. The frequency range of interest in measuring machinery noise is from about 30 Hz to 10 000 Hz.

GT noise characteristics

Gas turbines produce a wide variety of broad-band noise across the frequency range. There are three main emitters of noise: the machine's total surface, the air inlet system, and the exhaust gas outlet system. In practice, the inlet and outlet system noise is considered as included in the surface-originated noise. The machine bearings emit noise at frequencies related to their rotational speed, while the combustion process emits a wider, less predictable range of sound frequencies. Many industrial turbines are installed within an acoustic enclosure to reduce the levels of 'near vicinity' and environmental (further away) noise. Figure 9.27 shows the test arrangement.

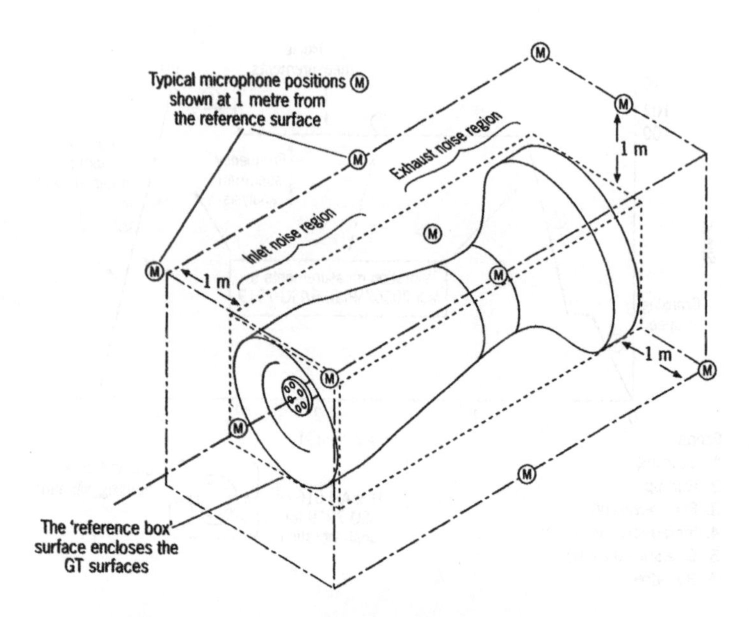

Commonly-used 'octave' mid-band frequencies are:

63 Hz	125 Hz	250 Hz	500 Hz	1000 Hz	2000 Hz	4000 Hz

Background noise correction (grade 2 accuracy ISO 10494)

Difference between 'running' and 'background' noise	*Subtract this correction from the 'running' noise*
6dB	1dB
7dB	1dB
8dB	1dB
9dB	1dB
10dB	0
>10dB	0

A noise test comprises the following steps:

1. Define the 'reference surface.'
2. Position the microphones 1 metre from the reference surface as shown.
3. Measure the background noise (at each microphone position)
4. Measure the 'running noise' at each microphone position – take A-weighted and discrete mid-band frequency readings.
5. Calculate the background noise correction – see the above table.
6. Remember that there is an 'uncertainty' tolerance of about ±2dB in most noise tests.

Fig. 9.27 GT noise tests

9.4 Gearboxes and testing

As precision items of rotating equipment, gearboxes are subject to various checks and tests during manufacture. The main checks during a test of a large spur, helical, or epicyclic gearbox are for:

- a correctly machined and aligned gear train;
- correctly balanced rotating parts;
- mechanical integrity of the components, particularly of the highly stressed rotating parts and their gear teeth.

Table 9.5 shows a typical acceptance guarantee schedule for a large gearbox.

Table 9.5 Large gearbox – typical acceptance guarantee schedule

The design standard	e.g. API 613
Rated input/output speeds	5200/3000 r/min
Overspeed capability	110 per cent (3300 r/min)
No-load power losses	Maximum 510 kW (this is sometimes expressed as a percentage value of the input power)
Oil flow	750 l/min (with a tolerance of ± 5 per cent)
Casing vibration	VDI 2056 group T: 2.8 mm/s r.m.s. (measured as a velocity)
Shaft vibration	Input pinion 39 μm
	Output shaft 50 μm peak-to-peak (both measured as an amplitude)
Noise level	ISO 3746: 97 dB(A) at 1 m distance

Gear inspection standards

Gear design and inspection standards are defined at the specification stage and relate to the application of the gearbox. Some commonly used ones are:

- **API 613**: (1988) *Special service gear units for refinery service*. This has direct relevance to works inspection and is used in many industries. For further technical details, API 613 cross-references the American Gear Manufacturers Association (AGMA) range of standards.
- **VDI 2056**: (1984) covers criteria for assessing mechanical vibration of machines. It is only applicable to the vibration of gearbox bearing housings and casings, not the shafts. Machinery is divided into six application 'groups' with gearboxes clearly defined as included in group

T. Vibration velocity (r.m.s.) is the measured parameter. Acceptance levels are clearly identified.

- **ISO 2372**: (1988) (equivalent to BS 4675) covers a similar scope to VDI 2056 but takes a different technical approach.
- **ISO 8579**: (1992) (equivalent to BS 7676) is in two parts, covering noise and vibration levels. It provides good coverage, but is not in such common use as other ISO and VDI standards.
- **ISO 7919/1**: (1986) (equivalent to BS 6749 Part 1). This relates specifically to the technique of measuring shaft vibration.
- **ISO 3746** and **API 615** are relevant noise standards. Other parts of test procedures addressed by standards are dynamic balancing and tooth contact tests.

Dynamic balancing test

Dynamic balancing is normally carried out after assembly of the gear wheels and pinions to their respective shafts. The rotor is spun at up to its rated speed and multiphase sensors, mounted in the bearings housings, sense the unbalance forces, relaying the values to a suitable instrument display. The purpose of dynamic balancing is to reduce the residual unbalance to a level which will ensure that the vibration characteristics of the assembled gearbox are acceptable. There is a useful first approximation for maximum permissible unbalance in API 613 (see Fig. 9.28). The correct compound unit is g.mm (gramme millimetres), i.e. an unbalance mass operating at an effective radius from the rotational axis. Any residual unbalance is corrected (after stopping the rotor) by adding weights into threaded holes. The test is then repeated to check the results.

Contact checks

Contact checks are a simple method of checking the meshing of a gear train. The results provide information about the machined accuracy of the gear teeth, and the relative alignment of the shafts. The test consists of applying a layer of 'engineers' blue' colour transfer compound to the teeth of one gear of each meshing set and then rotating the gears in mesh. The colour transfer shows the pattern of contact across each gear tooth (see Fig. 9.29).

Running tests

The mechanical running test is the key proving step for the gearbox. Most purchasers rely on a no-load running test, also referred to as a 'proof test'. The key objectives of the mechanical running test are to check that the oil flows, and that vibration and noise levels produced by the gearbox are

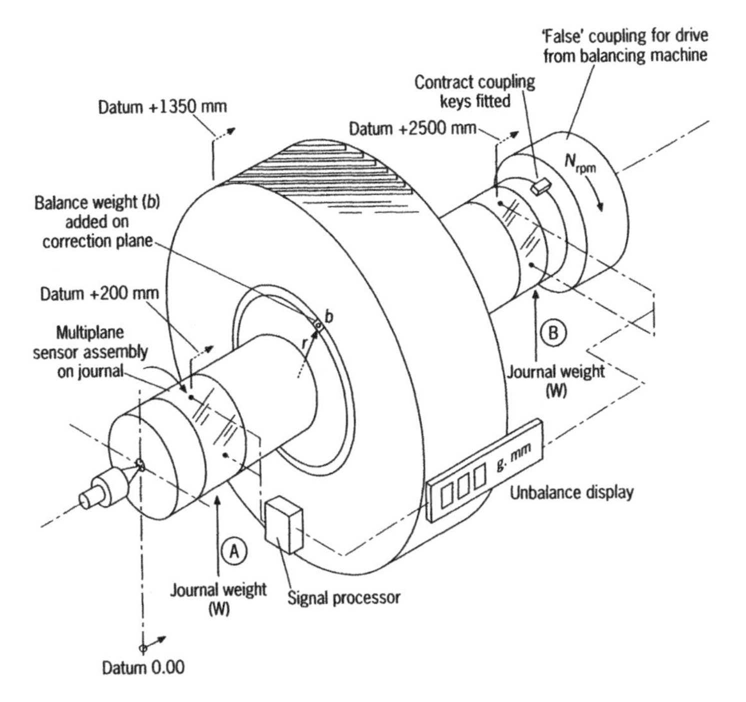

Record the results like this:

High-speed shaft. Rated (N) = rpm. Rotation anticlockwise (ACW) seen from coupling.

	Journal A	*Correction plane*	*Journal* B	
Location from datum 0.00	200mm	1350mm	2500mm	
Static loading (kg)	W	–	W	Specimen calculation (API 613) $\hat{U} = 6350\ W/N$
Balance wt added	–	b	–	
Residual unbalance (U) g.mm	U_A	–	U_B	

Fig. 9.28 Dynamic balance of a gear rotor

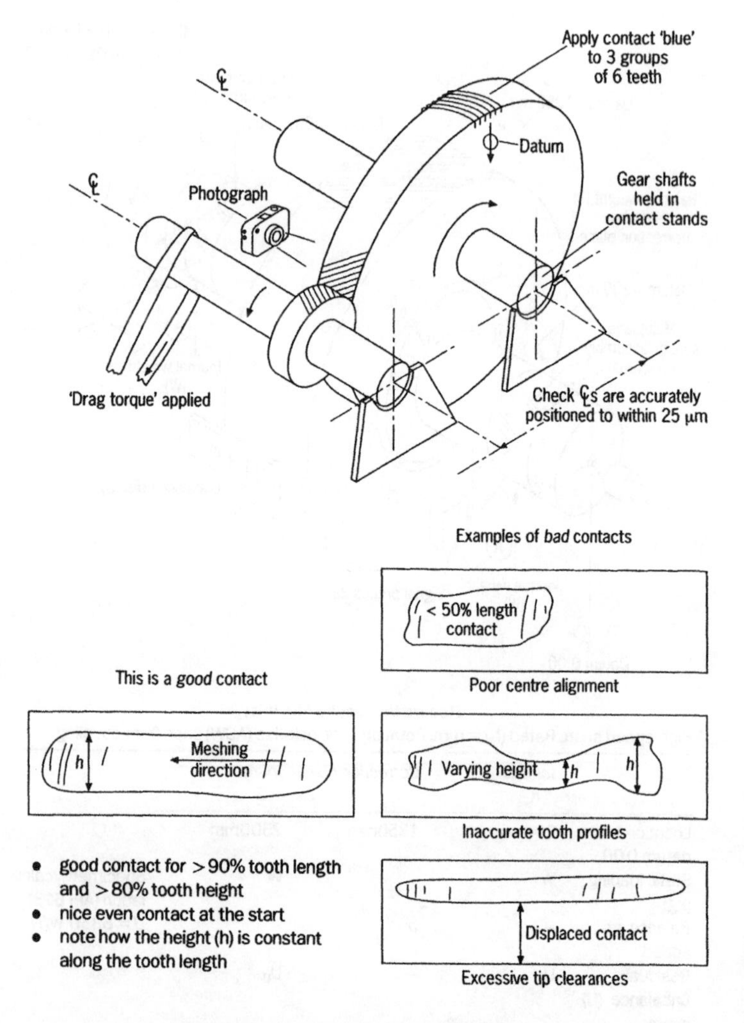

- good contact for >90% tooth length and >80% tooth height
- nice even contact at the start
- note how the height (h) is constant along the tooth length

Fig. 9.29 Gear train contact checks

within the guarantee acceptance levels. The test procedure contains the following steps (see Figs 9.30 and 9.31):

- slow-speed run
- run-up
- rated speed run
- overspeed test
- noise measurements
- stripdown.

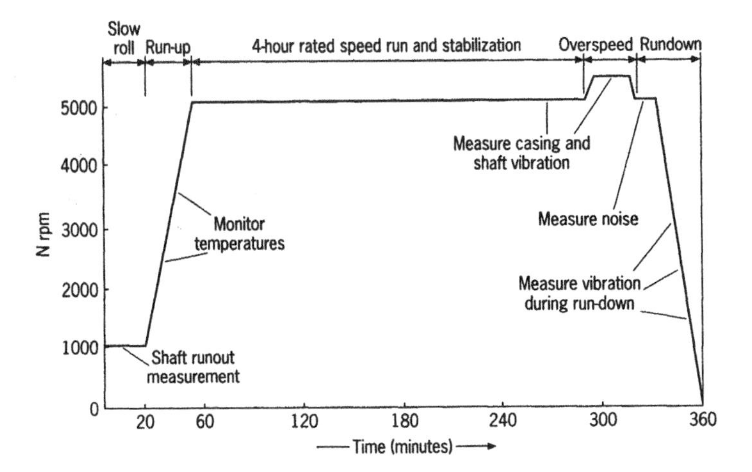

Fig. 9.30 Gearbox no-load running test

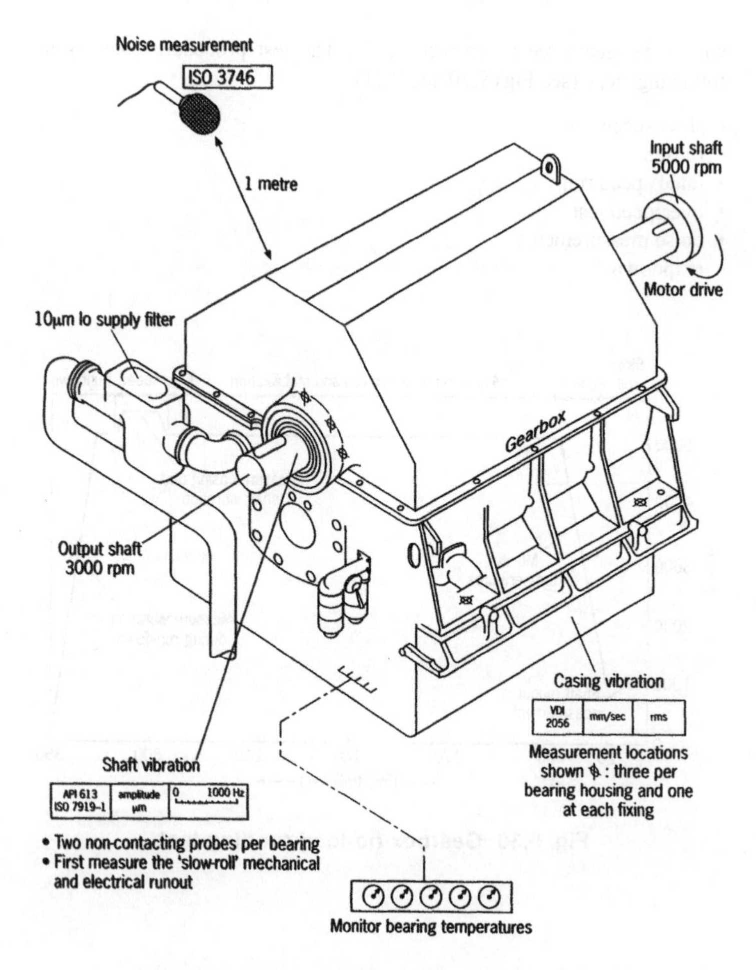

Fig. 9.31 Gearbox running test – monitoring

9.5 Reciprocating internal combustion engines

Internal combustion (IC) engines are classified basically into spark ignition (petrol) engines and compression ignition (diesel) engines. Petrol engines are used mainly for road vehicles up to a power of about 400 kW, while diesel engines, in addition to their use in road vehicles, are used in larger sizes for power generation, and locomotive and marine propulsion.

Diesel engines

Diesel engines are broadly divided into three categories based on speed. Table 9.6 gives a guide.

Table 9.6 Diesel engine speed categories

Designation	*Application*	*(Brake) Power rating (MW)*	*R/min*	*Piston speed (m/s)*
Slow speed (2 or 4 stroke)	Power generation, ship propulsion	Up to 45	<150	<9
Medium speed (4 stroke)	Power generation, ship propulsion	Up to 15	200–800	<12
High speed (4 stroke 'V')	Locomotives, portable power generation	Up to 5	>800	12–17

Diesel engine performance

Engine performance is the main issue that drives the ongoing design and development of diesel engines. The prime performance criterion is 'specific' fuel consumption (s.f.c.). Expressed in grammes (of fuel) per brake power (kW) hour, this is a direct way of assessing and comparing engines.

Engine guarantee schedules follow the pattern shown in Table 9.7, which shows indicative parameters for a medium-speed engine. Some key design requirements such as critical speed range, governor characteristics, acceptable bearing temperatures, and peak cylinder pressures are not stated explicitly, but do form part of an overall performance assessment of an engine. Figure 9.32 shows details of the brake test carried out to assess compliance with specified performance guarantees.

Table 9.7 Typical diesel engine guarantee schedule

GUARANTEES

- **Nominal site power output shall be 12MW (note the general statement here, to be qualified by reference to ISO 3046/2).**
- **Engine to be capable of 10% greater output for a period of one hour.**
- **Maximum continuous speed 500 rpm.**
- **Specified fuel: e.g. BS 2869 Class 'F'.**
- **Governing class: Type A1, single speed, accuracy class A1.**
- **The test standard shall be ISO 3046.**
- **Specific fuel consumption (units:grammes per (brake) kilowatt hour; g/(b)kWhr)**

MCR (%)	Guarantee (typical)	Guarantee +2.5%
50	220g/(b)kWhr	225g/(b)kWhr
60	216g/(b)kWhr	221g/(b)kWhr
80	214g/(b)kWhr	219g/(b)kWhr
90	212g/(b)kWhr	217g/(b)kWhr
100*	210g/(b)kWhr	205g/(b)kWhr

***Note that the 100% MCR specific fuel consumption guarantee is often qualified by a maximum allowable exhaust back-pressure.**

- **Emission levels**
  Maximum emission levels with the specified fuel grade at stated load (90%) shall not exceed:

NO_x	:	1400 mg/m^3
CO	:	450 mg/m^3
Particulates	:	100 mg/m^3
Non-methane hydrocarbons (NMH)	:	100 mg/m^3

- **Engine adjustment factors:**
  It is common to assess the engine design by evaluating closely the effect of retarding the fuel injection timing. You may see it expressed like this:

Effect of 2 degrees injection retard : Maximum +3% increase in sfc.
: Minimum decrease of -12% in NO_x
: Maximum increase of +3% in CO
: No measurable effect on Particulates or NMH.

***Remember*: These are good indicative figures but it depends on the particular engine design.**

- **Lubricating oil consumption**
  Guarantee l.o. consumption shall be less than 1.5 g/(b)kWhr (±5%) after site running-in for a period of 1000 running hours.
- **Site rating**
  At site conditions of : p = 1 bar (sea level), T = 300K
  relative humidity = 40 to 80%
  The rated power shall be 12000 kW.
  Site load condition will be 90% MCR.
  No percentage de-rating of the engine for site conditions.
- **Noise levels**
  At floor level and 1m distance from the engine the maximum allowable decibel level shall be a maximum of 115dB.

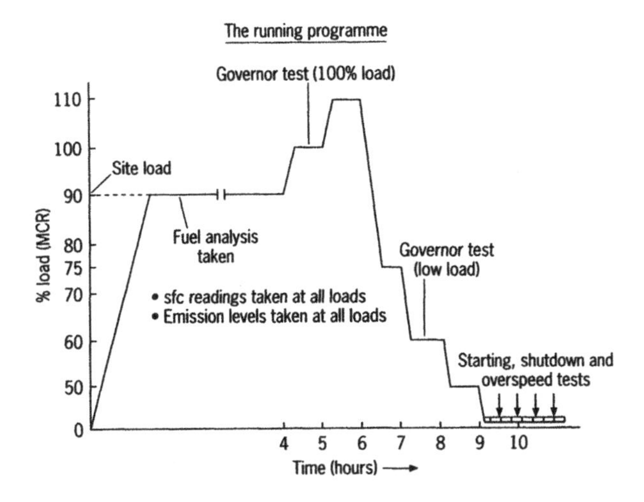

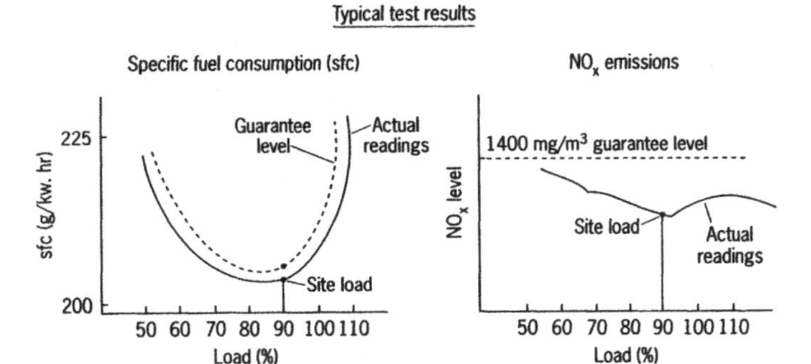

Fig. 9.32 Diesel engine brake test

Crankshaft deflections

Crankshaft deflection measurements are a way of checking crankshaft and bearing alignment. Figure 9.33 shows the basic methodology.

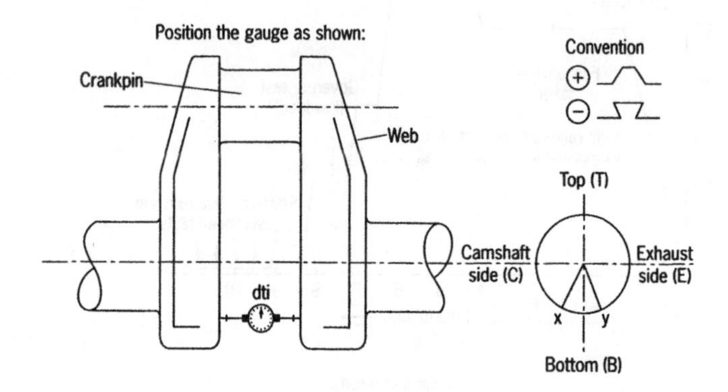

1. It is convention to set the gauge using x = 0 as a datum.
2. Don't confuse the algebra: e.g. (– 2) – (– 4) = + 2
3. Diagnosis. In this example the main bearing between cylinders 3 and 4 is high.

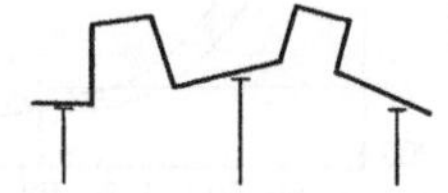

Typical readings (in 0.01mm) for a 6-cylinder engine

Crank position	Cylinder number					
	1	2	3	4	5	6
x	0	0	0	0	0	0
C	+4	+1	+3	-6	-2	+1
T	+8	+3	+10	-12	-6	+3
E	+4	+2	+5	-6	-4	+2
y	-2	+2	-2	0	0	-2
B=(x+y)/2	-1	+1	-1	0	0	-1
VM=T-B	+9	+2	+11	-12	-6	+4
HM=C-E	0	-1	-2	0	+2	-1

VM - Vertical misalignment
HM - Horizontal misalignment

Fig. 9.33 Crankshaft web deflections

Diesel engine technical standards

The main body of technical standards for diesel engines are those relating to testing of the engine. The most commonly used is ISO 3046: Parts 1 to 7 *Reciprocating internal combustion engines: performance*, identical in all respects to BS 5514 Parts 1 to 7.

- **ISO 3046/1** *Standard reference conditions* defines the standard 'ISO' reference conditions that qualify engine performance.
- **ISO 3046/2** *Test methods* describes the principles of acceptance guarantees during the load (brake) test and explains how to relate engine power at actual test conditions to ISO conditions and site conditions (both of which normally feature in the engine's guarantee specification).
- **ISO 3046/3** *Test measurements* gives permissible deviations for test temperatures and pressures.
- **ISO 3046/4** *Speed governing* defines the five possible classes of governing accuracy. An engine specification and acceptance guarantee will quote one of these classes.
- **ISO 3046/5** *Torsional vibrations* is a complex technical part of the standard and covers the principles of torsional vibration as applied specifically to reciprocating engines.
- **ISO 3046/6** *Specification of overspeed protection* defines the overspeed levels.
- **ISO 3046/7** *Codes for engine power* defines the way that manufacturers frequently classify the power output of their range of engines using a string of code numbers.

Table 9.8 shows some other related technical standards.

Table 9.8 Technical standards – IC engines

Standard	*Title*	*Status*
BS AU 141a: 1971	Specification for the performance of diesel engines for road vehicles.	Current
BS ISO 2697: 1999	Diesel engines. Fuel nozzles. Size 'S'.	Current
BS ISO 4093: 1999	Diesel engines. Fuel injection pumps. High-pressure pipes for testing.	Current
BS ISO 7299: 1996	Diesel engines. End-mounting flanges for fuel injection pumps.	Current
BS ISO 7612: 1994	Diesel engines. Base-mounted in-line fuel injection pumps. Mounting dimensions.	Current
BS EN ISO 8178-4: 1996	Reciprocating internal combustion engines. Exhaust emission measurement. Test cycles for different engine applications.	Current
BS EN ISO 8178-5: 1997	Reciprocating internal combustion engines. Exhaust emission measurement. Test fuels.	Current
BS ISO 14681: 1998	Diesel engines. Fuel injection pump testing. Calibrating fuel injectors.	Current
BS EN 1679 1: 1998	Reciprocating internal combustion engines. Safety. Compression ignition engines.	Current
96/714952	Reciprocating internal combustion engines. Performance. Part 4. Speed governing.	Current, draft for public comment

9.6 Turbochargers

Most modern diesel engines, and many petrol engines, are fitted with turbochargers. Because of their high speed, turbochargers are a highly complex and precision item of rotating equipment. Sizes range from large units fitted to marine diesel engines down to those used on motor vehicle engines. The smaller variants develop very quickly, as designs adapt to onerous conditions of speed, temperature, and fluid flow.

Principles of turbocharger operation

Figure 9.34 shows a small diesel engine turbocharger design. The exhaust gases of the engine are expanded in the tangential volute of the uncooled turbine casing and fed to a turbine rotor. Owing to the design of the volute, the gases are accelerated with a minimal loss of energy, so that the highest possible efficiency is attained. The turbine rotor drives the compressor impeller through the rotor shaft. The air to be compressed enters via a filter/silencer arrangement. After compression, the air is then fed to the engine inlet. A system of journal and roller bearings support the shaft, which rotates at high speed.

Turbochargers suffer from 'lag', a condition under which very little 'boost' air pressure is produced, owing to low exhaust gas volume flow when the engine is running at low speed. The overall design performance of a turbocharger is essentially controlled by three parameters: the type of exhaust turbine, the geometry ratio of the exhaust housing, and the geometry of the compressor.

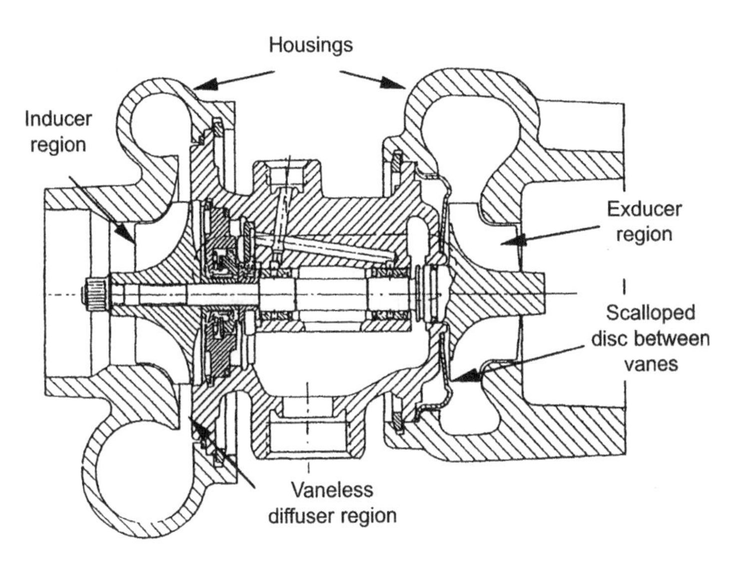

Fig. 9.34 Small turbocharger design

Exhaust turbine

Exhaust turbine design is a balance between absorbing as much energy from the exhaust gases as possible, and allowing the gases to flow as easily as possible. This is closely related to the size of the exhaust housing. A larger turbine can absorb more energy from the gases and spin the shaft with more torque and speed, but too large a turbine will restrict the flow of exhaust, thereby reducing engine performance.

Turbine housings geometry

Exhaust gas flow can be improved by incorporating design features such as nozzles or 'scrolls' that direct the exhaust gas flow directly onto the turbine blades. Such features affect the rate at which the turbine will accelerate, but these must be balanced against restricting the flow areas too much, since this will cause flow restrictions and increase undesirable backpressure on the engine, thereby reducing performance.

Compressor geometry

A turbocharger compressor housing is designed to convert the kinetic energy of the air into pressure energy. The size of the compressor turbine determines the maximum amount of air pressure 'boost' that the turbocharger can produce as well as its acceleration characteristic. In small engines the amount of boost produced is controlled by a 'wastegate'. The wastegate is a vacuum- or solenoid-actuated valve located at the exhaust inlet to the turbo which, when opened, causes the exhaust gases to bypass the exhaust turbine instead of passing through it. The further the wastegate is opened, the more exhaust is bypassed and the less boost is produced. Some designs also have a blow-off valve on the discharge manifold of the turbocharger casing. This is a vacuum-actuated valve that opens when there is vacuum in the intake manifold (closed-throttle). The release in pressure slows the run-down time of the turbocharger rotor and avoids undesirable pressure fluctuations.

CHAPTER 10

Draught Plant

10.1 Aeropropellers

A propeller, or 'airscrew', converts the torque of an engine (piston engine or turboprop) into thrust. Propeller blades have an airfoil section that becomes more 'circular' towards the hub. The torque of a rotating propeller imparts a rotational motion to the air flowing through it. Pressure is reduced in front of the blades and increased behind them, creating a rotating slipstream. Large masses of air pass through the propeller, but the velocity rise is small compared to that in turbojet and turbofan engines.

Blade element design theory

Basic design theory considers each section of the propeller as a rotating airfoil. The flow over the blade is assumed to be two dimensional (i.e. no radial component). From Fig. 10.1 the following equations can be expressed

Pitch angle $\phi = \tan^{-1}(V_o/\pi nd)$

u = velocity of blade element = $2\pi nr$

The propulsion efficiency (η_b) of the blade element, i.e. the 'blading efficiency', is defined by

$$\eta_b = \frac{V_o \mathrm{d}F}{u\mathrm{d}Q} = \frac{\tan\phi}{\tan(\phi+\gamma)} = \frac{L/D - \tan\phi}{L/D + \cot\phi}$$

Vector diagram for a blade element of a propeller

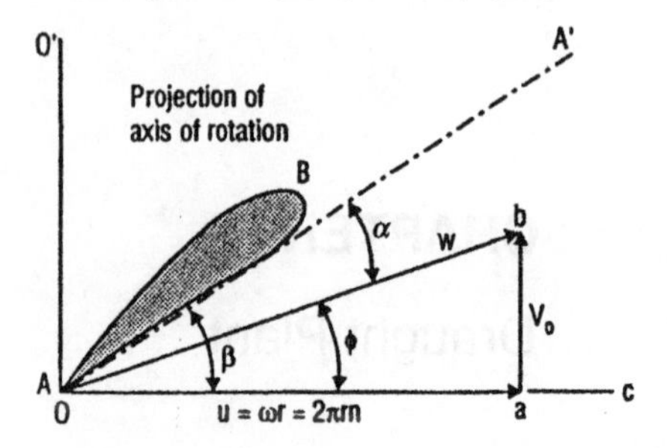

Aerodynamic forces acting on a blade element

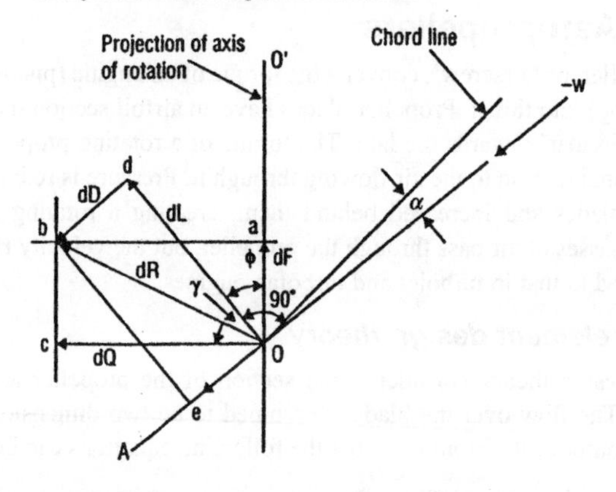

Fig. 10.1 Aeropropeller design

where

D = drag
L = lift
dF = thrust force acting on blade element
dQ = corresponding torque force
r = radius

The value of ϕ that makes η_b a maximum is termed the 'optimum advance angle' ϕ_{opt}.

Maximum blade efficiency is given by

$$(\eta_b)_{max} = \frac{2\gamma - 1}{2\gamma + 1} = \frac{2(L/D) - 1}{2(L/D) + 1}$$

Performance characteristics

The pitch and angle ϕ have different values at different radii along a propeller blade. It is common to refer to all parameters determining the overall characteristics of a propeller to their values at either $0.7r$ or $0.75r$.

Lift coefficient C_L is a linear function of the angle of attack α up to the point where the blade stalls, while drag coefficient C_D is a quadratic function of α. Figure 10.2 shows broad relationships between blading efficiency, pitch angle, and *L*/*D* ratio.

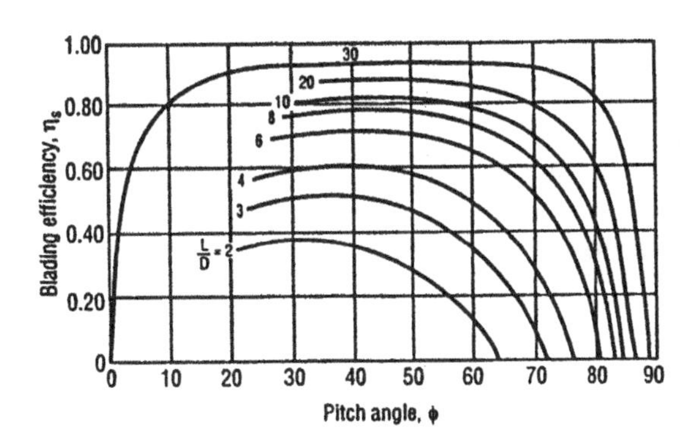

Fig. 10.2 A square key end shape

Propeller coefficients

It can be shown, neglecting the compressibility of the air, that

$$f(V_o, n, d_p, \rho, F) = 0$$

Using dimensional analysis, the following coefficients are obtained for expressing the performances of propellers having the same geometry

$$F = \rho n^2 d^4 p C_F \qquad Q = \rho n^2 d^5_p C_Q \qquad P = \rho n^3 d^5_p C_P$$

C_F, C_Q, and C_P are termed the thrust, torque, and power coefficients. These are normally expressed in USCS units

Thrust coefficient C_F

$$= \frac{F}{\rho n^2 d^4}$$

Torque coefficient C_Q

$$= \frac{Q}{\rho n^2 d^5}$$

Power coefficient C_P

$$= \frac{P}{\rho n^3 d^4}$$

where

d = propeller diameter (ft)
n = speed in revs/s
Q = torque (ft.lbs)
F = thrust (lbf)
P = power (ft.lbs/s)
ρ = air density (lb.s²/ft⁴)

Activity factor

Activity factor (AF) is a measure of the power-absorbing capabilities of a propeller, and hence a measure of its 'solidity'. It is defined as

$$AF = \frac{100\,000}{16} \int_{r_h/R}^{r/R=1} \frac{c}{d_P} \left(\frac{r}{R}\right)^3 d\left(\frac{r}{R}\right)$$

Propeller mechanical design

Propeller blades are subjected to:

- tensile stress due to centrifugal forces;
- steady bending stress due to thrust and torque forces;
- bending stress caused by vibration.

Vibration-induced stresses are the most serious, so propellers are designed so that their first-order, natural resonant frequency lies above expected operating speeds. To minimize the chance of failures, blades are designed using fatigue strength criteria. Steel blades are often hollow, whereas aluminium alloy ones are normally solid.

10.2 Draught fans

There are two main types of fan: axial and centrifugal. Axial fans are mainly used in low-pressure applications, making the centrifugal type the most common design. Figure 10.3 shows a typical large centrifugal fan.

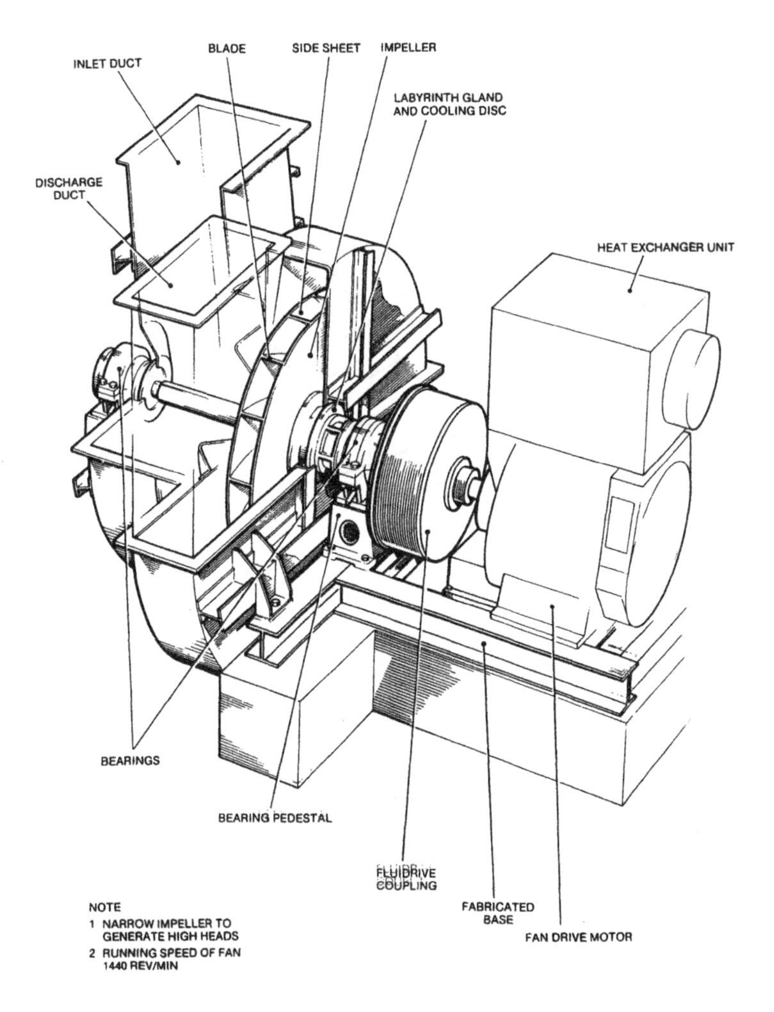

Fig. 10.3 Centrifugal draught fan – general view

Figure 10.4 shows a typical fan operating characteristic. Note how the characteristic reflects the amount of pressure it takes to push air through the system. Inlet control vanes are often used to help locate the operating point. The system resistance line shows how the fan efficiency reduces as the air flow and pressure decrease. An alternative method of control is by using variable speed drive motors or variable speed hydraulic couplings.

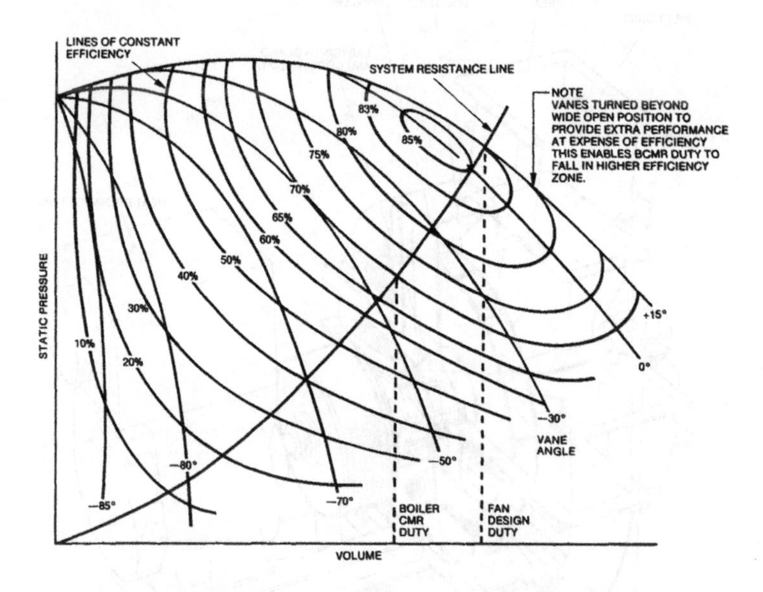

Fig. 10.4 Centrifugal fan – typical operating characteristic

Stall conditions

Every fan has a stable and unstable operating range. Stable flow is defined as the condition under which enough air flows through the fan wheel to provide a constant static pressure output for a given flow. As pressure is increased and flow reduced, the fan follows the curve to the left of the original operating point. At some point, the airflow is reduced to a point where individual fan blades, and then all the blades, enter a stalled condition, causing the flow regime to break down.

Critical speeds

Operation of a fan too near its critical speed will make it very sensitive to out-of-balance forces and resulting vibration. As a rule of thumb, the normal operating speed of a fan should be at least 20 per cent below the first critical speed.

Useful standards

Table 10.1 shows some published technical standards relating to draught fans and similar equipment.

Table 10.1 Technical standards – fans

Standard	*Title*	*Status*
BS 848-1: 1997, ISO 5801: 1997	Fans for general purposes. Performance testing using standardized airways.	Current
BS 848-2: 1985	Fans for general purposes. Methods of noise testing.	Current, partially replaced
BS 848-4: 1997, ISO 13351: 1996	Fans for general purposes. Dimensions.	Current
BS 848-5: 1999, ISO 12499: 1999	Fans for general purposes. Special for mechanical safety (guarding).	Current
BS 848-6: 1989	Fans for general purposes. Method of measurement of fan vibration.	Current, work in hand
BS 848-8: 1999, ISO 13349: 1999	Fans for general purposes. Vocabulary and definition of categories.	Current
BS 848-10: 1999, ISO 13350: 1999	Fans for general purposes. Performance testing of jet fans.	Current
BS 5060: 1987, IEC 60879: 1986	Specification for performance and construction of electric circulating fans and regulators.	Current, confirmed
BS EN 25136: 1994, ISO 5136: 1990	Acoustics. Determination of sound power radiated into a duct by fans. In-duct method.	Current, work in hand
BS EN 45510-4-3: 1999	Guide for the procurement of power station equipment. Boiler auxiliaries. Draught plant.	Current
88/72307 DC	General purpose industrial fans. Fan size designation (ISO/DIS 8171).	Current, draft for public comment

Table 10.1 Cont.

89/76909 DC	BS 848. Part 8. Fan terminology and classification.	Current, draft for public comment
95/704659 DC	Machines for underground mines. Safety requirements for mining ventilation machinery. Electrically driven fans for underground use (prEN 1872).	Current, draft for public comment
97/719334 DC	Ventilation for buildings. Air handling units. Ratings and performance for components and sections (prEN 13053).	Current, draft for public comment
98/704585 DC	Ventilation for buildings. Performance testing of components/products for residential ventilation. Part 4. Fans used in residential ventilation systems (prEN 13141-4).	Current, draft for public comment
98/718875 DC	Industrial fans. Performance testing in situ (ISO/DIS 5802).	Current, draft for public comment
00/561592 DC	Acoustics. Determination of sound power radiated into a duct by fans and other air-moving devices. In-duct method (ISO/DIS 5136).	Current, draft for public comment
00/704961 DC	ISO/DIS 14694. Industrial fans. Specification for balance quality and vibration levels.	Current, draft for public comment

10.3 'Fin-fan' coolers

Air-cooled, tube-nest heat exchangers (known loosely as 'fin-fan' coolers) are in common use for primary cooling purposes in desert areas or in inland plant sites. On a smaller scale, they have multiple uses in chemical and process plants where a self-contained cooling unit is needed, avoiding the complication of connecting every heat 'sink' component to a centralized cooling circuit. In their larger sizes, fin-fan coolers can cover an area of up to 4000–5000 m^2 and often stand up in a shallow angle 'A' configuration. Smaller ones usually stand horizontally, resting on a simple structural steel frame.

Construction

Figure 10.5 shows a basic fin-fan cooler design; they vary very little between manufacturers. The main design points are outlined below.

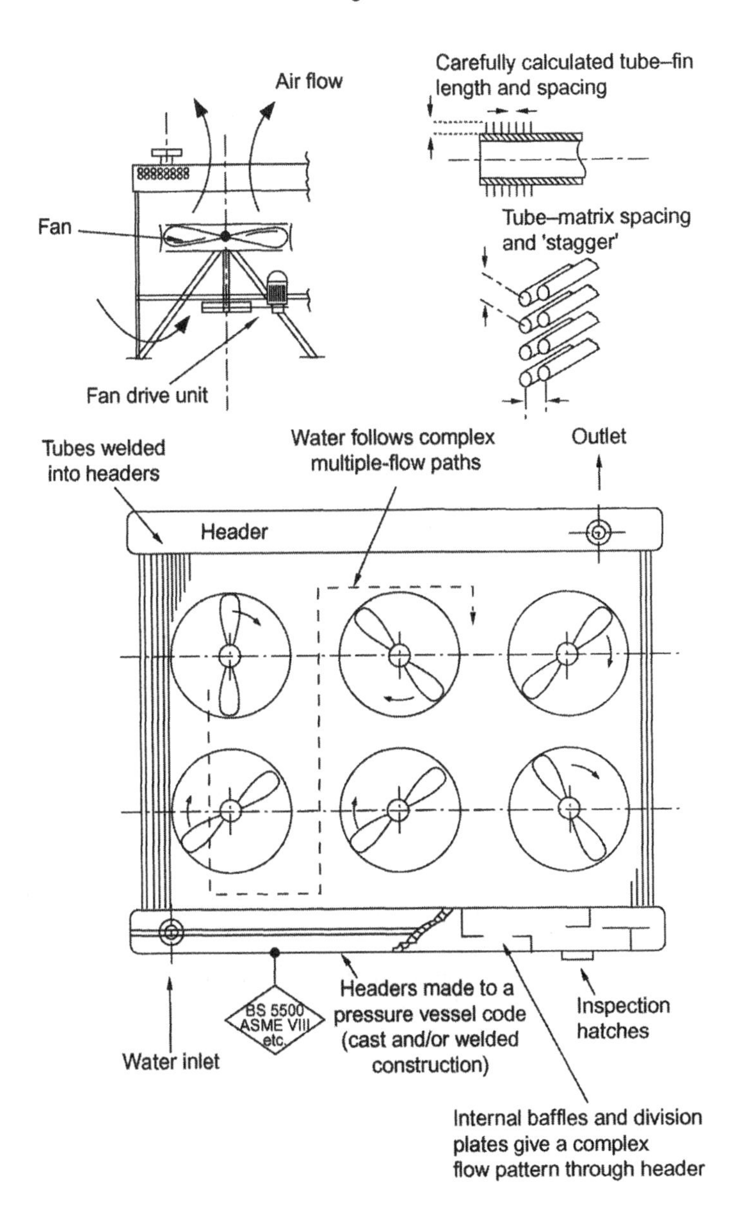

Fig. 10.5 'Fin-fan' cooler fan – general arrangement

The cooling matrix

This consists of a matrix of extruded carbon steel or stainless steel finned tubes arranged in a complex multi-pass flow path. The matrix is often divided into discrete banks of tubes, extending horizontally between a set of headers. The fins consist of a continuous spiral-wound, thin steel strip that is resistance-welded into a thin slot machined in a close helix around the tubes' outer surface. The extended surface of the fins adds significantly to the effective surface area, thereby increasing the overall thermal transfer. A typical tube bank is between six and ten tubes 'deep' in order to achieve the necessary heat transfer in as small a (horizontal) area as possible.

The headers

Each end of the tube banks are stub-welded into heavy-section cast and welded headers. These contain internal division plates and baffles that give the desired multi-pass pattern through the system. Each header also contains stub pieces and small access hatches for inspection, cleaning, and bleeding off unwanted air during commissioning. In most designs, the headers are designed and built to an accepted pressure vessel standard.

The air fans

Primary cooling effect is provided by a bank of axial-flow cooling fans that blow air vertically upwards through the tube nest. Fans are generally belt-driven for simplicity, and have variable incidence blades positioned by a pneumatic actuator arrangement. The electric motors are often two-speed (typically 300 r/min and 600 r/min), to allow operating current and power consumption to be reduced when air temperature is low. In a typical unit, each fan will be located about 2 m off the ground and will be protected by an expanded metal safety guard. Tip speed of the fan is normally kept below 60 m/s to avoid over-stressing the aluminium blades.

Fan running testing procedures

Fans are normally tested with their 'contract' motor – 'shop test' motors do not allow a proper assessment of the running current that will be experienced after site installation. Figure 10.6 shows a section through a typical fan, the shape of its performance characteristic, and the main points to check. The running test does not normally follow any particular technical standard; rather, it is organized around the task of demonstrating the fan's fitness-for-purpose in use. Specific points are as follows:

- *Static pressure versus blade angle* The performance of the fan does not keep on improving as blade incidence is increased. There is a well-defined 'cut-off point', above which the blades start to become aerodynamically inefficient and will actually produce less, rather than more, cooling effect.
- *Blade angle versus motor current* This places a limitation on the fitness-for-purpose of the fan. Maximum motor design currents usually have a design margin of about 30 per cent (to keep the cost of the motors down). A well-designed unit should reach full operating current before the static pressure curve levels off.
- *Vibration* Axial fans are normally smooth-running units and rarely experience vibration problems. A maximum V_{rms} level of about 2.5 mm/s is acceptable, using the principles of VDI 2056.
- *Mechanical integrity: points to check*
  - blade locking arrangements, including the fitted 'clevis', used to locate the blades accurately in position on the hub;
  - the pneumatic positioner and diaphragm that move the blade angle;
  - the blades themselves (usually aluminium): check for length and any obvious mechanical damage;
  - all locknuts and lockwashers fitted to the rotating components.

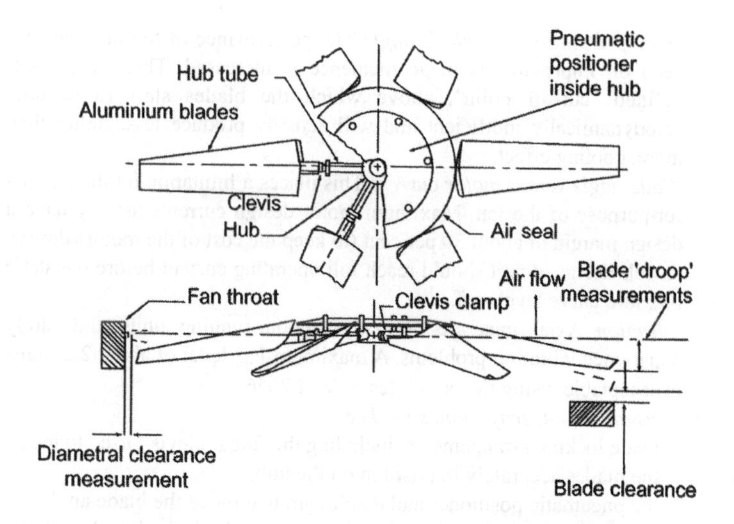

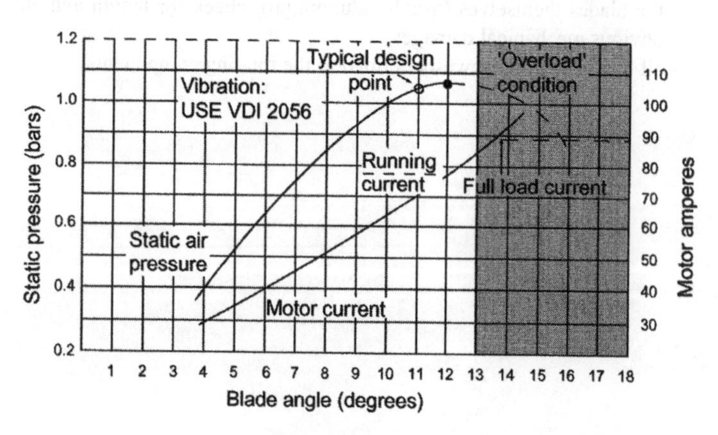

Fig. 10.6 'Fin-fan' cooler fan – typical performance characteristics

CHAPTER 11

Basic Mechanical Design

11.1 Engineering abbreviations

Table 11.1 shows abbreviations that are in common use in engineering drawings and specifications for rotating equipment.

Table 11.1 Engineering abbreviations

Abbreviation	*Meaning*
A/F	Across flats
ASSY	Assembly
CRS	Centres
L or CL	Centre line
CHAM	Chamfered
CSK	Countersunk
C'BORE	Counterbore
CYL	Cylinder or cylindrical
DIA	Diameter (in a note)
∅	Diameter (preceding a dimension)
DRG	Drawing
EXT	External
FIG.	Figure
HEX	Hexagon
INT	Internal
LH	Left hand
LG	Long
MATL	Material

Table 11.1 Cont.

MAX	Maximum
MIN	Minimum
NO.	Number
PATT NO.	Pattern number
PCD	Pitch circle diameter
RAD	Radius (in a note)
R	Radius (preceding a dimension)
REQD	Required
RH	Right hand
SCR	Screwed
SH	Sheet
SK	Sketch
SPEC	Specification
SQ	Square (in a note)
□	Square (preceding a dimension)
STD	Standard
VOL	Volume
WT	Weight

11.2 American terminology

In the USA, slightly different terminology is used, see Table 11.2. These abbreviations are based on the published standard ANSI/ASME Y14.5: (1994) *Dimensioning and tolerancing.*

Table 11.2 American abbreviations

Abbreviation	***Meaning***
ANSI	American National Standards Institute
ASA	American Standards Association
ASME	American Society of Mechanical Engineers
AVG	Average
CBORE	Counterbore
CDRILL	Counterdrill
CL	Centre line
CSK	Countersink

Table 11.2 Cont.

FIM	Full indicator movement
FIR	Full indicator reading
GD&T	Geometric dimensioning and tolerancing
ISO	International Standards Organisation
LMC	Least material condition
MAX	Maximum
MDD	Master dimension definition
MDS	Master dimension surface
MIN	Minimum
mm	Millimetre
MMC	Maximum material condition
PORM	Plus or minus
R	Radius
REF	Reference
REQD	Required
RFS	Regardless of feature size
SEP REQT	Separate requirement
SI	Système International (the metric system)
SR	Spherical radius
SURF	Surface
THRU	Through
TIR	Total indicator reading
TOL	Tolerance

11.3 Preferred numbers and preferred sizes

Preferred numbers are derived from geometric series in which each term is a uniform percentage larger than its predecessor. The first five principal series (named the 'R' series) are shown in Table 11.3.

Preferred numbers are taken as the basis for ranges of linear sizes of components, often being rounded up or down for convenience. Figure 11.1 shows the development of the R5 and R10 series.

Table 11.3 Preferred number series

Series	*Basis*	*Ratio of terms (% increase)*
R5	$5\sqrt{10}$	1.58 (58%)
R10	$10\sqrt{10}$	1.26 (26%)
R20	$20\sqrt{10}$	1.12 (12%)
R40	$40\sqrt{10}$	1.06 (6%)
R80	$80\sqrt{10}$	1.03 (3%)

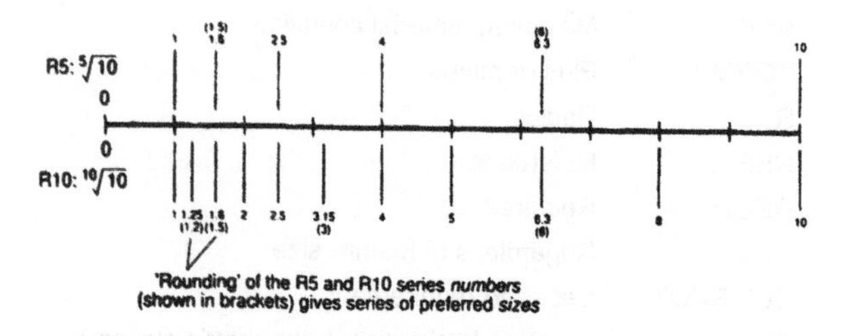

Fig. 11.1 The R5 and R10 series

Useful references

BS 2045: (1982) *Preferred numbers*. Equivalent to ISO 3.

11.4 Datums and tolerances – principles

A 'datum' is a reference point or surface from which all other dimensions of a component are taken; these other dimensions are said to be 'referred to' the datum. In most practical designs, a datum surface is usually used, this generally being one of the surfaces of the machine element itself rather than an 'imaginary' surface. This means that the datum surface normally plays an important part in the operation of the elements. The datum surface is usually machined and may be a mating surface or a locating face between elements, or similar (see Figs 11.2 and 11.3). Simple machine mechanisms do not always need datums – it depends on what the elements do and how complicated the mechanism assembly is.

A 'tolerance' is the allowable variation of a linear or angular dimension about its 'perfect' value. British Standard 308 and similar published standards contain accepted methods and symbols.

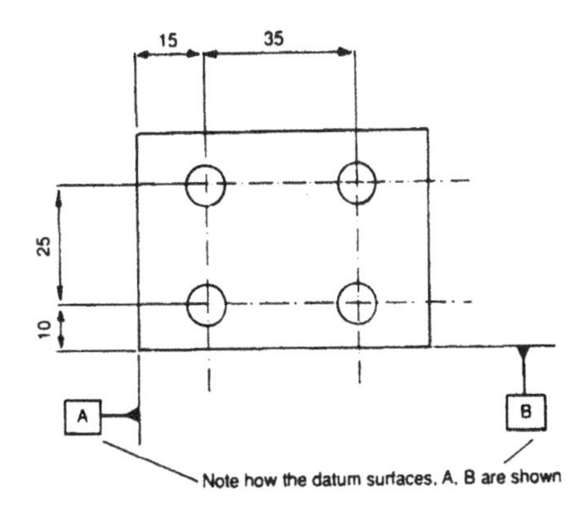

Fig. 11.2 Datums and tolerances

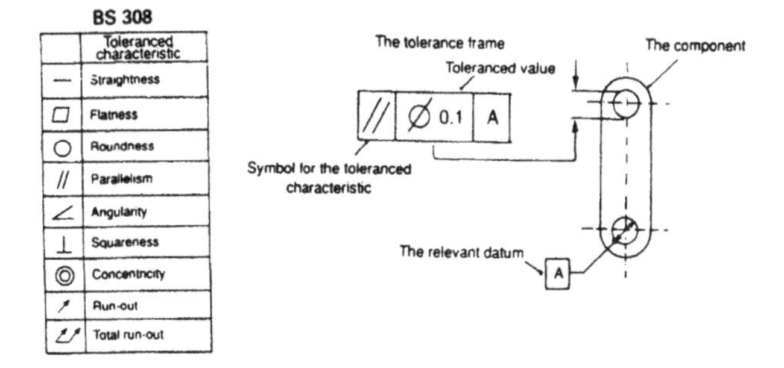

Fig. 11.3 The tolerance frame

Toleranced dimensions

In designing any engineering component it is necessary to decide which dimensions will be toleranced. This is predominantly an exercise in necessity – only those dimensions that *must* be tightly controlled, to preserve the functionality of the component, should be toleranced (see Fig. 11.4). Too many toleranced dimensions will increase significantly the manufacturing costs and may result in 'tolerance clash', where a dimension derived from other toleranced dimensions can have several contradictory values.

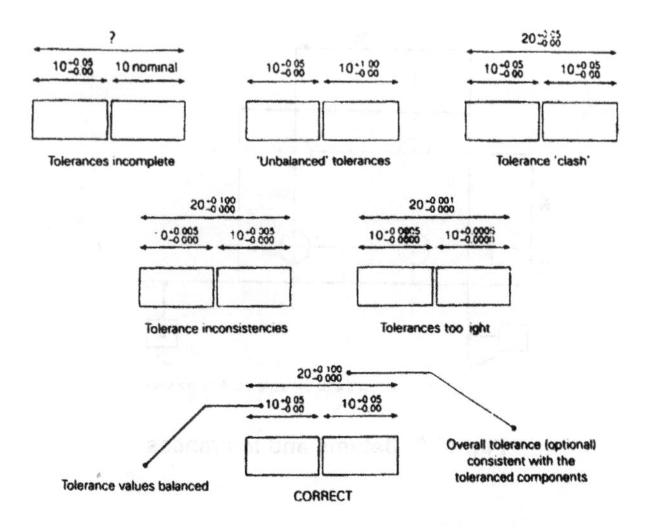

Fig. 11.4 Toleranced dimensions

General tolerances

It is a sound principle of engineering practice that in any rotating machine design there will only be a small number of toleranced features. The remainder of the dimensions will not be critical.

There are two ways to deal with this. First, an engineering drawing or sketch can be annotated to specify that a 'general tolerance' should apply to features where no specific tolerance is mentioned. This is often expressed as ± 0.5 mm. Alternatively, the drawing can make reference to a 'general tolerance' standard such as BS EN 22768, which gives typical tolerances for linear dimensions, as shown in Table 11.4.

Table 11.4 Typical tolerances for linear dimensions

Dimension (mm)	*Tolerance (mm)*
0.6–6.0	± 0.1
6–36	± 0.2
36–120	± 0.3
120–315	± 0.5
315–1000	± 0.8

11.5 Holes

The tolerancing of holes depends on whether they are made in thin sheet (up to about 3 mm thick) or in thicker plate material. In thin material, only two toleranced dimensions are required (see Fig. 11.5).

- *Size* A toleranced diameter of the hole, showing the maximum and minimum allowable dimensions.
- *Position* Position can be located with reference to a datum and/or its spacing from an adjacent hole. Holes are generally spaced by reference to their centres.

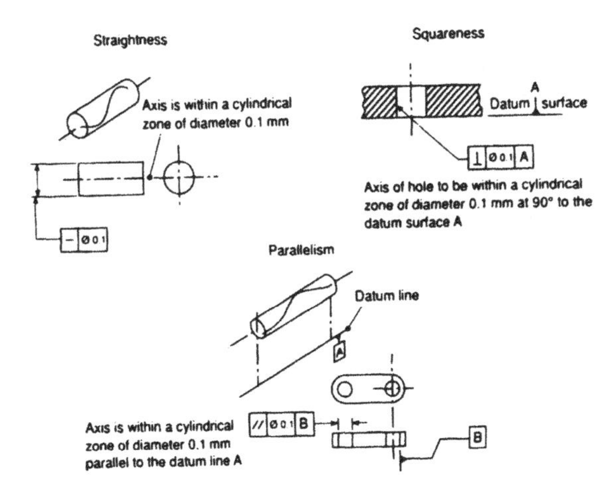

Fig. 11.5 Tolerances – holes

For thicker material, three further toleranced dimensions become relevant: straightness, parallelism, and squareness.

- *Straightness* A hole or shaft can be straight without being perpendicular to the surface of the material.
- *Parallelism* This is particularly relevant to holes and is important when there is mating hole-to-shaft fit.
- *Squareness* The formal term for this is 'perpendicularity'. Simplistically, it refers to the squareness of the axis of a hole to the datum surface of the material through which the hole is made.

11.6 Screw threads

There is a well-established system of tolerancing adopted by British and International Standard Organizations and the manufacturing industry. This system uses the two complementary elements of fundamental deviation and tolerance range to define fully the tolerance of a single component. It can be applied easily to components, such as screw threads, which join or mate together (see Fig. 11.6).

- *Fundamental deviation (FD)* is the distance (or 'deviation') of the nearest 'end' of the tolerance band from the nominal or 'basic' size of a dimension.
- *Tolerance band (or 'range')* is the size of the tolerance band, i.e. the difference between the maximum and minimum acceptable size of a toleranced dimension. The size of the tolerance band, and the location of the FD, governs the system of limits and fits applied to mating parts.

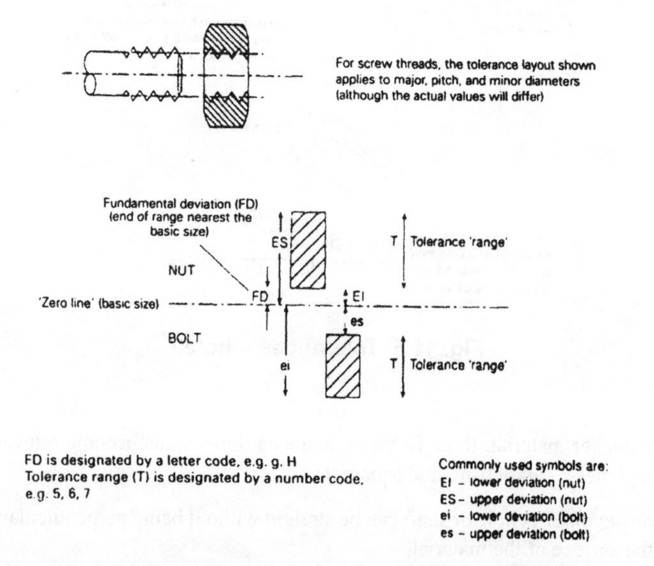
For screw threads, the tolerance layout shown applies to major, pitch, and minor diameters (although the actual values will differ)
Fundamental deviation (FD) (end of range nearest the basic size)
NUT
ES
FD
T
Tolerance 'range'
'Zero line' (basic size)
EI
es
BOLT
ei
T
Tolerance 'range'
FD is designated by a letter code, e.g. g. H
Tolerance range (T) is designated by a number code, e.g. 5, 6, 7
Commonly used symbols are:
EI – lower deviation (nut)
ES – upper deviation (nut)
ei – lower deviation (bolt)
es – upper deviation (bolt)

Fig. 11.6 Tolerances – screw threads

Tolerance values have a key influence on the costs of a manufactured item, so their choice must be seen in terms of economics as well as engineering practicality. Mass-produced items are competitive and price sensitive, and over-tolerancing can affect the economics of a product range.

11.7 Limits and fits

Principles

In machine element design there is a variety of different ways in which a shaft and hole are required to fit together. Elements such as bearings, location pins, pegs, spindles, and axles are typical examples. The shaft may be required to be a tight fit in the hole, or to be looser, giving a clearance to allow easy removal or rotation. The system designed to establish a series of useful fits between shafts and holes is termed 'limits and fits'. This involves a series of tolerance grades so that machine elements can be made with the correct degree of accuracy and can be interchangeable with others of the same tolerance grade (see Fig. 11.7).

The British Standard BS 4500/BS EN 20286 *ISO Limits and fits* contains the recommended tolerances for a wide range of engineering requirements. Each tolerance grade is designated by a combination of letters and numbers, such as IT7, which would be referred to as grade 7.

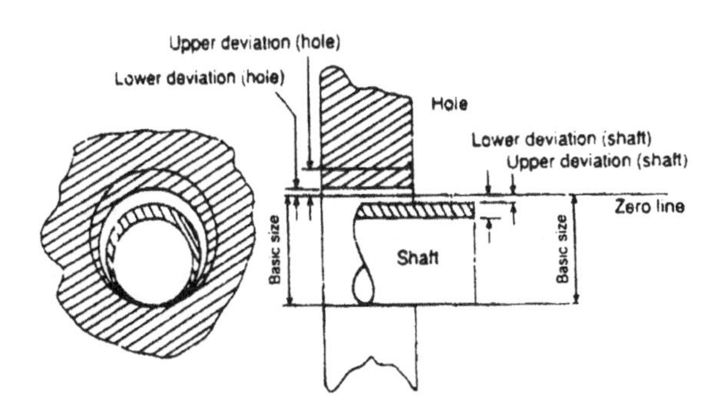

Fig. 11.7 Limits and fits

Figure 11.7 shows the principles of a shaft/hole fit. The 'zero line' indicates the basic or 'nominal' size of the hole and shaft (it is the same for each) and the two shaded areas depict the tolerance zones within which the hole and shaft may vary. The hole is conventionally shown above the zero line. The algebraic difference between the basic size of a shaft or hole and its actual size is known as the 'deviation'.

- It is the deviation that determines the nature of the fit between a hole and a shaft.
- If the deviation is small, the tolerance range will be near the basic size, giving a tight fit.
- A large deviation gives a loose fit.

Various grades of deviation are designated by letters, similar to the system of numbers used for the tolerance ranges. Shaft deviations are denoted by small letters, and hole deviations by capital letters. Most general engineering uses a 'hole-based' fit in which the larger part of the available tolerance is allocated to the hole (because it is more difficult to make an accurate hole) and then the shaft is made to suit, to achieve the desired fit.

Common combinations

There are seven popular combinations used in general mechanical engineering design (see Fig. 11.8).

1. *Easy running fit*: H11–c11, H9–d10, H9–e9. These are used for bearings where a significant clearance is necessary.
2. *Close running fit*: H8–f7, H8–g6. This only allows a small clearance, suitable for sliding spigot fits and infrequently used journal bearings. This fit is not suitable for continuously rotating bearings.
3. *Sliding fit*: H7–h6. Normally used as a locational fit in which close-fitting items slide together. It incorporates a very small clearance and can still be freely assembled and disassembled.
4. *Push fit*: H7–k6. This is a transition fit, mid-way between fits that have a guaranteed clearance and those where there is metal interference. It is used where accurate location is required, e.g. dowel and bearing inner-race fixings.
5. *Drive fit*: H7–n6. This is a tighter grade of transition fit than the H7–k6. It gives a tight assembly fit where the hole and shaft may need to be pressed together.
6. *Light press fit*: H7–p6. This is used where a hole and shaft need permanent, accurate assembly. The parts need pressing together but the fit is not so tight that it will overstress the hole bore.

7. *Press fit*: H7–s6. This is the tightest practical fit for machine elements such as bearing bushes. Larger interference fits are possible but are only suitable for large, heavy, engineering components.

Holes: H11, H9, H9, H8, H7, H7, H7, k6, H7, n6, H7, p6, H7, s6

Shafts: c11, d10, e9, f7, g6, h6

Nominal size in mm	Clearance fits												Transition fits				Interference fits			
	Easy running						Close running				Sliding		Push		Drive		Light press		Press	
	Tols*		Tols		Tols		Tols		Tols		Tols		Tols		Tols		Tols		Tols	
	H11	c11	H9	d10	H9	e9	H8	f7	H7	g6	H7	h6	H7	k6	H7	n6	H7	p6	H7	s6
6–10	+90 0	−80 −170	+36 0	−40 −98	+36 0	−25 −61	+22 0	−12 −28	+15 0	−5 −14	+15 0	−9 0	+15 0	+10 +1	+15 0	+19 +10	+15 0	+24 +15	+15 0	+32 +23
10–18	+110 0	−95 −205	+43 0	−50 −120	+43 0	−32 −75	+27 0	−16 −34	+18 0	−6 −17	+18 0	−11 0	+18 0	+12 +1	+18 0	+23 +12	+18 0	+29 +18	+18 0	+39 +28
18–30	+130 0	−110 −240	+52 0	−65 −149	+52 0	−40 −92	+33 0	−20 −41	+21 0	−7 −20	+21 0	−13 0	+21 0	+15 +2	+21 0	+28 +15	+21 0	+35 +22	+21 0	+48 +35
30–40	+140 0	−120 −280	+62	−80	+62	−50	+39	−25	+25	−9	+25	−16	+25	+18	+25	+33	+25	+42	+25	+59
40–50	+160 0	−130 −290	0	−180	0	−112	0	−50	0	−25	0	0	0	+2	0	+17	0	+26	0	+43

* Tolerance units in 0.001 mm

Data from BS 4500

Fig. 11.8 Limits and fits – common combinations

11.8 Surface finish

Surface finish, more correctly termed 'surface texture', is important for all machine elements that are produced by machining processes such as turning, grinding, shaping, or honing. This applies to surfaces that are flat or cylindrical. Surface texture is covered by its own technical standard, BS 1134 *Assessment of surface texture*. It is measured using the parameter R_a, which is a measurement of the average distance between the median line of the surface profile and its peaks and troughs, measured in micrometres (μm). There is another system from a comparable standard, DIN ISO 1302, which uses a system of 'N' numbers – it is simply a different way of describing the same thing (see Fig. 11.9).

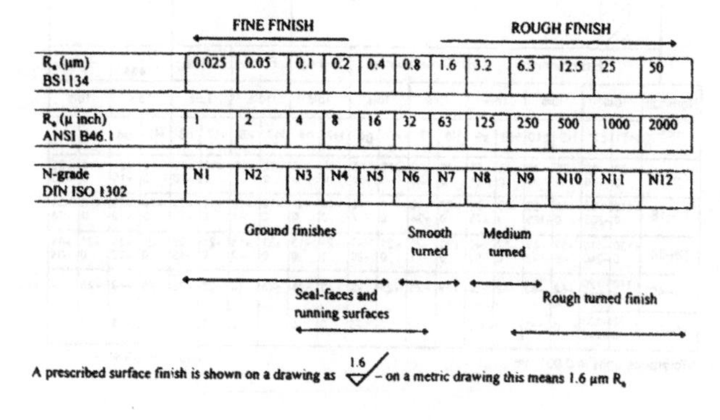

Fig. 11.9 Surface finish

Choice of surface finish: 'rules of thumb'

- Rough turned, with visible tool marks: N10 (12.5 μm R_a)
- Smooth machined surface: N8 (3.2 μm R_a)
- Static mating surfaces (or datums): N7 (1.6 μm R_a)
- Bearing surfaces: N6 (0.8 μm R_a)
- Fine 'lapped' surfaces: N1 (0.025 μm R_a)

Finer finishes can be produced but are more suited for precision applications such as instruments. It is good practice to specify the surface finish of close-fitting surfaces of machine elements, as well as other BS 308 parameters such as squareness and parallelism.

11.9 Reliability in design

The concept of reliability is an important consideration of rotating equipment design. There is a well-developed theoretical side to it: quantities such as MTTF (mean time to failure) and MTBF (mean time between failures) are in common use in safety-critical applications such as petroleum and chemical plant design. In essence:

- **RELIABILITY IS ABOUT *HOW*, *WHY*, AND *WHEN* THINGS FAIL.**

The 'bathtub curve'

This is so-called for no better reason than it looks, in outline, like a bathtub (see Fig. 11.10). It indicates when you can expect things to fail and is well proven, reflecting reasonably accurately what happens to many engineering products. It tends to be most accurate for complex products, including most rotating equipment. The chances of failure are quite high in the early operational life of a product item. This is due to inherent defects or fundamental design errors in the product, or incorrect assembly of the multiple component parts. A progressive wear regime then takes over for the middle 75 per cent of the product's life – the probability of failure here is low. As the lifetime progresses, the rate of deterioration increases, causing progressively higher chances of failure.

Failure mode analysis

Failure mode analysis (FMA) is concerned with how and why failures occur. In contrast to the bathtub curve there is a strong product-specific bias to this technique, so generalized 'results' rarely have much validity. In theory, most engineered products will have a large number of possible ways that they can fail (termed 'failure modes'). Practically, this reduces to three or four common types of failure, because of particular design parameters, distribution of stress, or similar. The technique of FMA is a structured look at all the possibilities, so that frequently-occurring failure modes can be anticipated in advance of their occurring, and can be 'designed out'. FMA is therefore, by definition, multidisciplinary. Figure 11.11 outlines the principles of FMA, using as an example a simple compression spring – a common sub-component of many rotating equipment products.

Risk analysis

This encapsulates a number of assessment techniques that are all 'probabilistic'. They look at a failure in terms of the probability that it may, or may not, happen. The techniques tend to be robust in the mathematical sense, but sometimes have rather limited practical application because the

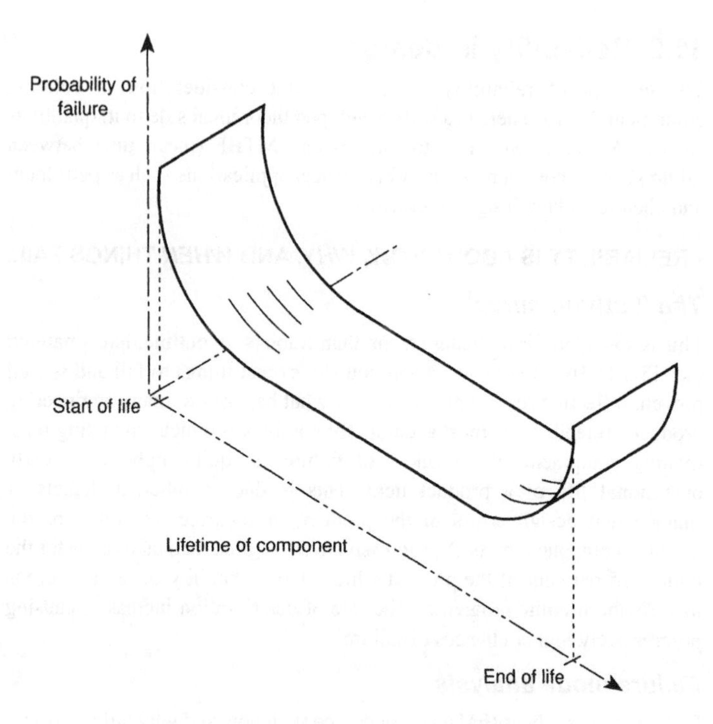

Fig. 11.10 Reliability – the 'bathtub curve'

rules of probability are not axiomatic in the engineering world. This means that risk analysis techniques are fine, as long as you realize the limitations of their use when dealing with practical engineering designs.

Reliability assessment

The most useful form of reliability assessment involves looking forward, to try and eliminate problems before they occur, to effect reliability improvements. There are limitations to this technique:

- it involves anticipation, which is difficult;
- reliability assessment is strictly relative. It may be possible to conclude that component X should last longer than component Y, but not that component X will definitely last for 50 000 h;
- it must be combined with sound engineering and design knowledge if it is to be effective.

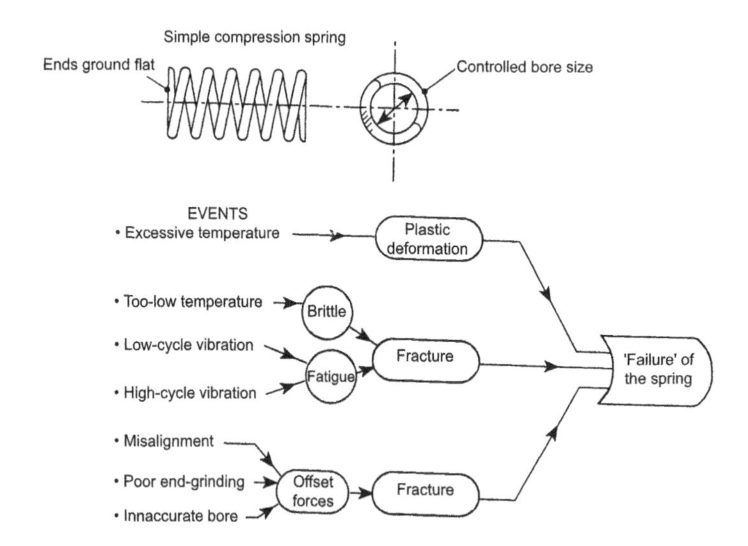

Fig. 11.11 The principles of failure mode analysis (FMA)

11.10 Improving design reliability: eight principles

Reduce static loadings

It is good practice to reduce static loadings on component parts, by redistributing loads or increasing loaded areas. The effects are small when existing design stresses are less than about 30 per cent of yield strength (R_e), but can be significant if they are higher. The amount of deformation of a component is reduced, which can lead to a decreased incidence of failure. Low stresses improve reliability.

Reduce dynamic loadings

In many components, stresses caused by dynamic loadings can be several orders of magnitude higher than 'static' design stresses – up to nine or ten times if shock loads are involved. Dynamic shock loads are, therefore, a major cause of failure. It is good design practice to eliminate as many shock loads as possible. This can be done by using design features such as damping, movement restricters, flexible materials, and by isolating critical components from specific externally induced shock loadings. The possibility of general dynamic stresses can normally be limited by reducing the relative speed of movement of components. In rotating shafts, particularly, this has the effect of reducing internal torsional stresses during starting and braking of the shaft.

Reduce cyclic effects

Cyclic fatigue is the biggest cause of failure of engineering components. High-speed, low-speed, and normally static components frequently fail in this way. The mechanism is well known; cyclic stresses as low as 40 per cent of R_e will cause progressive weakening of most materials. One of the major principles of improving reliability therefore is to reduce cyclic effects wherever possible. This applies to the amplitude of the loading and to its frequency. Typical cyclic effects are:

- *vibration*: this is defined in three orthogonal planes x, y, and z; it is often caused by residual unbalance of rotating parts;
- *pulsations*: often caused by pressure fluctuations;
- *twisting*: in many designs, torsion is cyclic, rather than static; heavy duty pump shafts and engine crankshafts are good examples;
- *deflections*: designs that have members which are intended to deflect in use invariably suffer from cyclic fatigue to some degree. In applications such as aircraft structures, fatigue life is the prime criterion that determines the useful life of the product.

Reduce operating temperatures

This applies to the majority of moving components that operate at above-ambient temperature. Excessively high temperatures, for instance in bearings and similar 'contact' components, can easily cause failure. The principle of reliability improvement is to increase the design margin between the operating temperature of a bearing face or lubricant film and its maximum allowable temperature. Typical actions include increasing cooling capacity and lubricant flow, or reducing specific loadings. The effect of temperature on static components is also an important consideration. Thermal expansion of components with complex geometry can be difficult to calculate accurately and can lead to unforeseen deflections, movements, and 'locked in' assembly stresses. Stresses induced by thermal expansion of constrained components can be extremely high, capable of fracturing most engineering materials quite easily. For low-temperature components (normally static, such as aircraft external parts or cryogenic pressure vessels) the problem is the opposite; low temperatures increase the brittleness (decrease the Charpy impact resistance) of most materials. This is hard to 'design out' as low temperatures are more often the result of a component's environment, rather than its actual design. As a general principle, aim for component design temperatures as near to ambient as possible – it helps improve reliability.

Remove stress-raisers

Stress-raisers are sharp corners, grooves, notches, or acute changes of section that cause stress concentrations under normal loadings. They can be found on both rotating and static components. The stress concentration factors of sharp corners and grooves are high, and difficult to determine accurately. Components that have failed predominantly by a fatigue mechanism are nearly always found to exhibit a 'crack initiation point' – a sharp feature at which the crack has started and then progressed by a cyclic fatigue mechanism to failure. There are techniques that can be employed to reduce stress-raisers:

- use blended radii instead of sharp corners, particularly in brittle components such as castings;
- for rotating components like drive shafts, keep rotating diameters as constant as possible. If it is essential to vary the shaft diameter, use a taper. Avoid sharp shoulders, grooves, keyways, and slots;
- avoid rough surface finishes on rotating components. A rough finish can act as a significant stress-raiser – the surface of a component is often furthest from its neutral axis and therefore subject to the highest level of stress.

Reduce friction

Although friction is an essential part of many engineering designs, notably machines, it inevitably causes wear. It is good practice, therefore, to aim to reduce non-essential friction, using good lubrication practice and/or low-friction materials whenever possible. Lubrication practice is perhaps the most important one; aspects such as lubricating fluid circuit design, filtering, flowrates, and flow characteristics can all have an effect on reliability. If you can keep friction under controlled conditions, you will improve reliability.

Design for accurate assembly

Large numbers of engineering components and machines fail because they are not assembled properly. Precision rotating machines such as engines, gearboxes, and turbines have closely specified running clearances and cannot tolerate much misalignment in assembly. This applies even more to smaller components such as bearings, couplings, and seals. Reliability can be improved, therefore, by designing a component so that it can only be assembled accurately. This means using design features such as keys, locating lugs, splines, guides, and locating pins which help parts assemble together accurately. It is also useful to incorporate additional measures to make it impossible to assemble components the wrong way round or back-to-front. Accurate assembly can definitely improve reliability (although you will not find a mathematical theory explaining why).

Isolate corrosive and erosive effects

As a general principle, corrosive and erosive conditions, whether from a process fluid or the environment, are detrimental to most materials in some way. They cause failures. It is best to keep them isolated from close-fitting moving parts – the use of clean flushing water for slurry pump bearings and shaft seals is a good example. Corrosion and erosion also attack large, unprotected static surfaces, so highly resistant alloys or rubber/epoxy linings are often required. Galvanic corrosion is an important issue for small closely matched component parts of machines – look carefully at the electrochemical series for the materials being used to see which one will corrode sacrificially. Good design reliability is about making sure the less critical components corrode first. It is sometimes possible to change the properties of the electrolyte (often process or flushing fluid, or oil) to reduce its conductivity, if a potential difference exists between close-fitting components and galvanic corrosion problems are expected.

11.11 Design for reliability – a new approach

Design for reliability (DFR) is an evolving method of stating and evaluating design issues in a way that helps achieve maximum reliability in a design. The features of this 'new approach' are:

- it is a quantitative but visual method – so not too difficult to understand;
- no separate distinction is made between the functional performance of a design and its reliability – both are considered equally important;
- it does not rely on pre-existing failure rate data (which can be inaccurate).

The technique

Design parameters are chosen with the objective of maximizing all of the safety margins that will be built in to a product or system. All the possible modes of failure are investigated and then expressed as a set of design constraints (see Fig. 11.12). The idea is that a design which has the highest safety margin with respect to all the constraints will be the most reliable (point X in the figure). Constraints are inevitably defined in a variety of units, so a grading technique is required that yields a non-dimensional performance measure of each individual constraint.

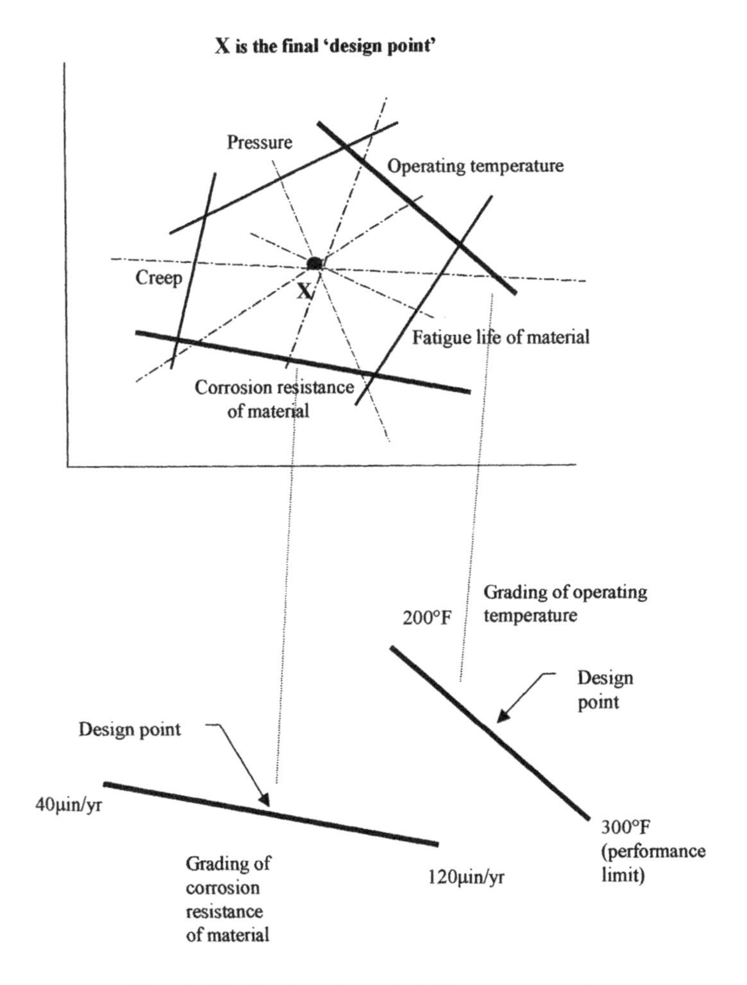

Fig. 11.12 Design for reliability – expressing design constraints

11.12 Useful references and standards

1. BS 308 *Engineering drawing practice* (various parts).
2. BS EN 22768-1: (1993) *Working limits on toleranced dimensions.*
3. BS 4500/BS EN 20286: (1993) *ISO system of limits and fits* (various parts).
4. DIN ISO 1302: (1992) *Technical drawings – methods of indicating surface texture.*
5. DIN ISO 1101: (1983) *Technical drawings.*
6. DIN ISO 8015: (1985) *Technical drawings – fundamental tolerancing principles.*
7. DIN 4768: (1983) *Surface roughness.*
8. BS 1134: *Assessment of surface texture* (various parts).
9. ANSI/ASME Y14.5M: (1994) *Dimensioning and tolerancing.*
10. ISO 286-1: (1988) *ISO system of limits and fits.*

CHAPTER 12

Materials of Construction

Material properties are of great importance in all aspects of rotating equipment design and manufacture. It is essential to check the up-to-date version of the relevant British Standards or equivalent when choosing or assessing a material. The most common materials used for rotating equipment are divided into the generic categories of carbon, alloy, stainless steel, and non-ferrous.

12.1 Plain carbon steels – basic data

Typical properties are shown in Table 12.1.

Table 12.1 Plain carbon steel: properties

Type	*%C*	*%Mn*	*Yield, R_e (MN/m^2)*	*UTS, R_m (MN/m^2)*
Low C steel	0.1	0.35	220	320
General structural steel	0.2	1.4	350	515
Steel castings	0.3	–	270	490

12.2 Alloy steels – basic data

Alloy steels have various amounts of Ni, Cr, Mn, or Mo added to improve properties. Typical properties are shown in Table 12.2.

Table 12.2 Alloy steels: properties

Type	*%C*	*Others (%)*	*R_e (MN/m²)*	*R_m (MN/m²)*
Ni/Mn steel	0.4	0.85 Mn 1.00 Ni	480	680
Ni/Cr steel	0.3	0.5 Mn 2.8 Ni 1.0 Cr	800	910
Ni/Cr/Mo steel	0.4	0.5 Mn 1.5 Ni 1.1 Cr 0.3 Mo	950	1050

12.3 Stainless steels – basic data

Stainless steel is a generic term used to describe a family of steel alloys containing more than about 11 per cent chromium. The family consists of four main classes, subdivided into about 100 grades and variants. The main classes are austenitic and duplex. The other two classes, ferritic and martensitic, tend to have more specialized application and so are not so commonly found in general rotating equipment use. The basic characteristics of each class are given below.

- *Austenitic* The most commonly used basic grades of stainless steel are usually austenitic. They have 17–25 per cent Cr, combined with 8–20 per cent Ni, Mn, and other trace alloying elements which encourage the formation of austenite. They have low carbon content, which makes them weldable. They have the highest general corrosion resistance of the family of stainless steels.
- *Ferritic* Ferritic stainless steels have high chromium content (>17 per cent Cr) coupled with medium carbon, which gives them good corrosion resistance properties rather than high strength. They normally have some Mo and Si, which encourage the ferrite to form. They are generally non-hardenable.
- *Martensitic* This is a high-carbon (up to 2 per cent C), low-chromium (12 per cent Cr) variant. The high carbon content can make it difficult to weld.
- *Duplex* Duplex stainless steels have a structure containing both austenitic and ferritic phases. They can have a tensile strength of up to twice that of straight austenitic stainless steels and are alloyed with various trace elements to aid corrosion resistance. In general, they are as weldable as austenitic grades but have a maximum temperature limit, because of the characteristic of their microstructure.

Table 12.3 gives basic stainless steel data.

Table 12.3 Stainless steels – basic data

Stainless steels are commonly referred to by their AISI equivalent classification (where appropriate).

AISI	Other classifications	Type +	Yield F_{ty} (ksi)	[(R_e) MPa]	Ultimate F_{tu} (ksi)	[(R_m) MPa]	E(%) 50 mm	HRB	%C	%Cr	% others *	Properties
302	ASTM A296 (cast), Wk 1.4300, 18/8, SIS 2331	Austenitic	40	[275.8]	90	[620.6]	55	85	0.15	17–19	8–10 Ni	A general purpose stainless steel.
304	ASTM A296, Wk 1.4301, 18/8/LC, SIS 2333, 304S18	Austenitic	42	[289.6]	84	[579.2]	55	80	0.08	18–20	8–12 Ni	An economy grade.
304L	ASTM A351, Wk 1.4306, 18/8/ELC, SIS 2352, 304S14	Austenitic	39	[268.9]	80	[551.6]	55	79	0.03	18–20	8–12 Ni	Low C to avoid intercrystalline corrosion after welding.
316	ASTM A296, Wk 1.4436, 18/8/Mo, SIS 2243, 316S18	Austenitic	42	[289.6]	84	[579.2]	50	79	0.08	16–18	10–14 Ni	Addition of Mo increases corrosion resistance.
316L	ASTM A351, Wk 1.4435, 18/8/Mo/ELC, 316S14, SIS 2353	Austenitic	42	[289.6]	81	[558.5]	50	79	0.03	16–18	10–14 Ni	Low C weldable variant of 316.

Table 12.3 Cont.

321	ASTM A240, Wk 1.4541, 18/8/Ti, SIS 2337, 321S18	Austenitic	35	[241.3]	90	[620.6]	45	80	0.08	17–19	9–12 Ni	Variation of 304 with Ti added to improve temperature resistance.
405	ASTM A240/A276/ A351, UNS 40500	Ferritic	40	[275.8]	70	[482.7]	30	81	0.08	11.5–14.5	1 Mn	A general purpose ferritic stainless steel.
430	ASTM A176/A240/ A276, UNS 43000, Wk 1.4016	Ferritic	50	[344.7]	75	[517.1]	30	83	0.12	14–18	1 Mn	Non-hardening grade with good acid-resistance.
403	UNS S40300, ASTM A176/A276	Martensitic	40	[275.8]	75	[517.1]	35	82	0.15	11.5–13	0.5 Si	Turbine grade of stainless steel.
410	UNS S40300, ASTM A176/A240, Wk 1.4006	Martensitic	40	[275.8]	75	[517.1]	35	82	0.15	11.5–13.5	4.5–6.5 Ni	Used for machine parts, pump shafts, etc.
–	255 (Ferralium)	Duplex	94	[648.1]	115	[793]	25	280 HV	0.04	24–27	4.5–6.5 Ni	Better resistance to SCC than 316.
–	Avesta SAF 2507 §, UNS S32750	'Super' duplex 40% ferrite	99	[682.6]	116	[799.8]	~ 25	300 HV	0.02	25	7 Ni, 4 Mo, 0.3 N	High strength. Max. temp 575 °F (301 °C) due to embrittlement.

* Main constituents only shown.
+ All austenitic grades are non-magnetic; ferritic and martensitic grades are magnetic.
§ Avesta trade mark

12.4 Non-ferrous alloys – basic data

The term 'non-ferrous alloys' is used for those alloy materials that do not have iron as their base element. The main ones used for mechanical engineering applications, with their ultimate tensile strength ranges, are:

- nickel alloys 400–1200 MN/m^2
- zinc alloys 200–360 MN/m^2
- copper alloys 200–1100 MN/m^2
- aluminium alloys 100–500 MN/m^2
- magnesium alloys 150–340 MN/m^2
- titanium alloys 400–1500 MN/m^2

The main ones in use are nickel alloys, in which nickel is frequently alloyed with copper or chromium and iron to produce material with high temperature and corrosion resistance. Typical types and properties are shown in Table 12.4.

Table 12.4 Nickel alloys: properties

Alloy type	*Designation*	*Constituents (%)*	*UTS (MN/m^2)*
Ni–Cu	UNS N04400 ('Monel')	66 Ni, 31 Cu, 1 Fe, 1 Mn	415
Ni–Fe	'Ni lo 36'	36 Ni, 64 Fe	490
Ni–Cr	'Inconel 600'	76 Ni, 15 Cr, 8 Fe	600
Ni–Cr	'Inconel 625'	61 Ni, 21 Cr, 2 Fe, 9 Mo, 3 Nb	800
Ni–Cr	'Hastelloy C276'	57 Ni, 15 Cr, 6 Fe, 1 Co, 16 Mo, 4 W	750
Ni–Cr (age hardenable)	'Nimonic 80A'	76 Ni, 20 Cr	800–1200
Ni–Cr (age hardenable)	'Inco Waspalloy'	58 Ni, 19 Cr, 13 Co, 4 Mo, 3 Ti, 1 Al	800–1000

12.5 Material traceability

The issue of material traceability is an important aspect of the manufacture of high-integrity rotating equipment. Most technical codes and standards make provision for quality assurance activities designed to ensure that materials of construction used in the pressure envelope are traceable.

Figure 12.1 shows the 'chain of traceability' which operates for rotating equipment materials. Note that although all the activities shown are

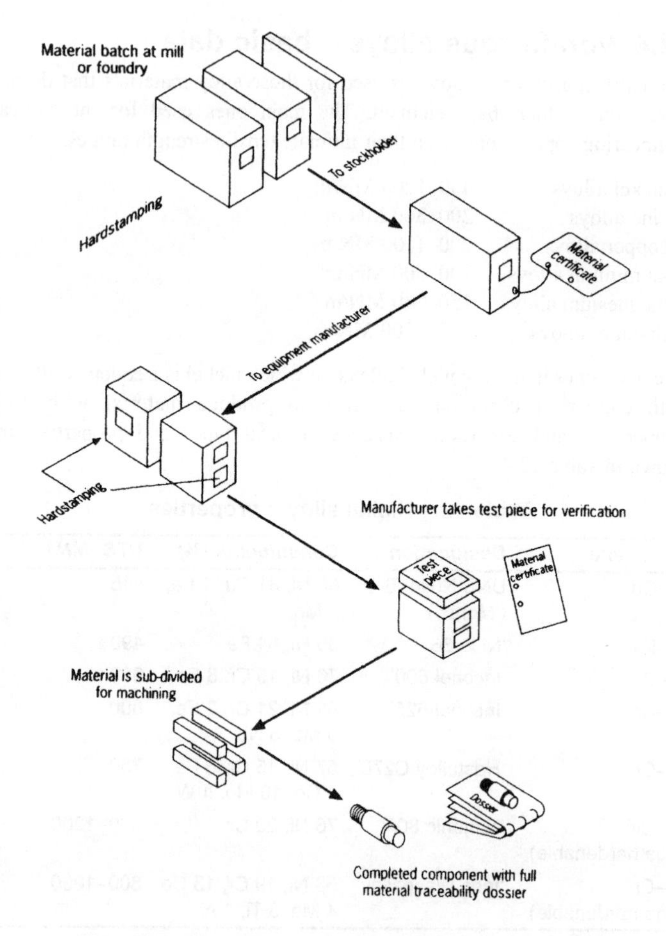

Fig. 12.1 The 'chain of traceability' for materials

available for use (i.e. to be specified and then implemented) this does not represent a unique system of traceability suitable for all materials. In practice there are several 'levels' in use, depending on both the type of material and the nature of its final application. The most common document referenced in the material sections of rotating equipment specifications is the European Standard EN 10 204: (1991) *Metallic products – types of inspection documents*. It provides for two main 'levels' of certification: Class '3' and Class '2' (see Table 12.5).

Table 12.5 Material traceability: EN 10 204 classes

EN 10 204 certificate type	*Document validation by*	*Compliance with:*		*Test results included*	*Test basis*	
		the order	*'technical rules' **		*Specific*	*Non-specific*
3.1A	I	•	•	Yes	•	–
3.1B	M(Q)	•	•	Yes	•	–
3.1C	P	•		Yes	•	
3.2	P + M(Q)	•		Yes	•	–
2.3	M			Yes	•	–
2.2	M			Yes	–	•
2.1	M	•	–	No	–	•

I – An independent (third party) inspection organization.

P – The purchaser.

M(Q) – An 'independent' (normally QA) part of the material manufacturer's organization.

M – An involved part of the material manufacturer's organization.

* – Normally the 'technical rules' on material properties given in the relevant material standard (and any applicable technical code).

CHAPTER 13

The Machinery Directives

13.1 The Machinery Directive 98/37/EC – what is it?

The Machinery Directive 98/37/EC is a prominent European 'new approach' directive with major implications for manufacturers and importers of all types of machinery and components, including most rotating equipment. The current directive 98/37/EC has evolved from various previous directives (including 89/392/EEC, 98/368/EEC, and 93/44/EEC) and, as such, represents a consolidation of the content of these earlier directives. The Machinery Directive takes its place as one of the family of New Approach directives that impose identical requirements in every member state within the European Economic Area (EEA).

In line with all European directives, The Machinery Directive has to be implemented in each member state by national regulations. In the UK, this is Statutory Instrument SI 1992/3073 as amended by SI 1994/2063: *The supply of machinery (safety) regulations 1994* (amended). These are enforced by the Health and Safety Executive (HSE) for machinery used in the workplace.

13.2 New Approach directives

The concept of New Approach directives was introduced in 1985 as a move towards a coherent family of directives that follow a particular construction – the overall objective being eventual harmonization of technical 'rules' across European member states. Because of the difficulty in rationalizing the technical content of different published technical codes and standards, new approach directives do not mention specific technical standards but, instead, contain a set of 'essential safety requirements' (ESRs) with which all products covered by the directive must comply. In addition, the objectives of the ESRs are supported by the content of European Harmonized Standards.

So, the two alternative ways of complying with a new approach directive (such as The Machinery Directive 98/37/EC) are:

- compliance with a relevant European Harmonized Standard; this provides a presumption of conformity with the ESRs in the directive;
- compliance with other, non-harmonized national standards (or no published standards at all) plus confirmation that the ESRs have been met in some other way.

There is a further all-encompassing requirement that products must be accompanied by all the appropriate documentation and must, in fact, be safe.

13.3 The scope of The Machinery Directive

The Directive applies to all machinery and 'safety components'. A definition is that a machine is defined as 'an assembly of linked parts or components, at least one of which moves'.

There are some exclusions from The Directive:

- machines that are already covered by other directives (see Table 13.1);
- equipment that falls within the scope of the Low Voltage Directive 73/23/EEC and 93/68/EEC.

Table 13.1 Exclusions from The Machinery Directive

The following categories of machines are excluded from the jurisdiction of The Machinery Directive. This is because they are low risk, or because they are of high risk and, therefore, covered by other directives or requirements (see The Machinery Directive). They are:

1 Machinery whose only power source is directly applied manual effort, unless it is a machine used for lifting or lowering loads.

2 Machinery for medical use, used in direct contact with patients.

3 Special equipment for use in fairgrounds and/or amusement parks.

4 Steam boilers, tanks, and pressure vessels.

5 Machinery specially designed or put into service for nuclear purposes which, in the event of failure, may result in an emission of radioactivity.

6 Radioactive sources forming part of a machine.

7 Firearms.

Table 13.1 Cont.

8	Storage tanks and pipelines for petrol, diesel fuel, inflammable liquids, and dangerous substances.
9	Means of transport, i.e. vehicles and their trailers intended solely for transporting passengers by air or on road, rail, or water networks, as well as means of transport insofar as such means are designed for transporting goods by air, on public road or rail networks, or on water. (Vehicles used in the mineral extraction industry are not excluded.)
10	Sea-going vessels and mobile offshore units together with equipment on board such vessels or units.
11	Cableways, including funicular railways, for public or private transport of persons.
12	Agricultural and forestry tractors.
13	Machines specially designed and constructed for military or police purposes.
14	Lifts that permanently serve specific levels of buildings and constructions, having a car moving between guides which are rigid and inclined at an angle of more than 15 degrees to the horizontal and designed for the transport of: – persons; – persons and goods; – goods alone if the car is accessible (a person may enter it without difficulty) and fitted with controls situated inside the car or within reach of a person inside.
15	Means of transport of persons using rack and pinion rail mounted vehicles.
16	Mine winding gear.
17	Theatre elevators.
18	Construction site hoists intended for lifting persons or persons and goods.

Old machinery manufactured before 1995 does not have to comply, unless it is refurbished or upgraded to the extent that its specification is substantially changed. Some specialized machines are identified by The Machinery Directive as requiring special test procedures. These are shown in Table 13.2.

Table 13.2 Types of machinery and safety components subject to 'special attestation procedures'

A. MACHINERY

1 Circular saws (single or multi-blade) for working with wood and analogous materials and working with meat and analogous materials:
- sawing machines with fixed tool during operation, having fixed bed with manual feed of the workpiece or with a demountable power feed;
- sawing machines with fixed tool during operation, having a manually operated reciprocating saw-bench or carriage;
- sawing machines with fixed tool during operation, having a built-in mechanical feed device for the workpieces, with manual loading and/or unloading;
- sawing machines with moveable tool during operation, with a mechanical feed device and manual loading and/or unloading.

2 Hand-fed surface planing machines for woodworking.

3 Thicknessers for one-side dressing with manual loading and/or unloading for woodworking.

4 Band-saws with fixed or mobile bed and band saws with a mobile carriage, with manual loading and/or unloading, for working with wood and analogous materials and for working with meat and analogous materials.

5 Combined machines of the types referred to in 1–4 and 7 for working with wood and analogous materials.

6 Hand-fed tenoning machines with several tool holders for woodworking.

7 Hand-fed vertical spindle moulding machines for working with wood and analogous materials.

8 Portable chain saws for woodworking.

9 Presses, including press-brakes, for the cold working of metals, with manual loading and/or unloading, whose movable working parts may have a travel exceeding 6 mm and a speed exceeding 30 mm/s.

10 Injection or compression plastics-moulding machines with manual loading or unloading.

11 Injection or compression rubber-moulding machines with manual loading or unloading.

12 Machinery for underground working of the following types:
- machinery on rails: locomotives and brake-vans;
- hydraulic powered roof supports;
- internal combustion engines to be fitted to machinery for underground working.

Table 13.2 Cont.

13	Manually loaded trucks for the collection of household refuse incorporating a compression mechanism.
14	Guards and detachable transmission shafts with universal joints as described in Section 3.4.7 of Annex B.
15	Vehicles servicing lifts.
16	Devices for the lifting of persons involving a risk of falling from a vertical height of more than three metres.
17	Machines for the manufacture of pyrotechnics.
B.	***SAFETY COMPONENTS***
1	Electro-sensitive devices designed specifically to detect persons in order to ensure their safety (non-material barriers, sensor mats, electro-magnetic detectors, etc.).
2	Logic units that ensure the safety functions of bi-manual controls.
3	Automatic moveable screens to protect the presses referred to in 9, 10, and 11.
4	Roll-over protection structures (ROPS).
5	Falling-object protective structures (FOPS).

Goals, principles, and structures

The overall goals and principles of The Machinery Directive are complex, but the main points are:

- The manufacturer (or his 'authorized representative' in any member state) is responsible for the machine being safe.
- Design must be based on 'state of the art' and 'good practice' and decisions taken during the design and development phases need to be properly documented to demonstrate this.
- The effect of The Directive, as implemented by statutory instrument in member states, has most effect on the way that a machinery manufacturer deals with engineering design development, manufacturing, and documentation practice rather than the precise form of the end product (the machine itself).
- Instead of specifying prescriptive technical requirements, The Directive requires that machines meet a set of essential safety requirements (ESRs). Most of the ESRs have some direct relation to minimizing safety hazards that may be present.

The structure of The Machinery Directive consists of a series of formal 'articles' followed by six main 'annexes'. These annexes are the most important content from a design and engineering viewpoint. Table 13.3 summarizes their content.

Table 13.3 The Machinery Directive annexes

Annex I	This contains the essential safety requirements (ESRs). A general part applies to all machinery and specific parts apply specifically to certain categories of machine.
Annex II	This specifies the need for the issue of a certificate or declaration of conformity that must be provided with every machine supplied for 'independent use' (i.e. for sale).
Annex III	Specifies the requirement and method of CE marking to signify compliance with the directive.
Annex IV	This is a list of machinery and safety components that are subject to special testing procedures that have to be carried out in conjunction with a Notified Body.
Annex V	This describes the EC declaration of conformity and shows the content of the 'technical file'.
Annex VI	This relates to the criteria for selection, by national authorities, of Notified Bodies.

13.4 The CE mark – what is it?

The CE mark (see Fig. 13.1) probably stands for Communiteé Européen. Unfortunately, it is far from certain that whoever invented the mark (probably a bureaucrat in Brussels) had anything particular in mind, other than to create a logo that would be universally recognized in the European Union. It is best thought of as simply a convenient logo, without any deeper meaning. In its current context, a machine can only have the CE mark fixed to it if it complies with all European directives pertaining to that type of product, including, if applicable, The Machinery Directive.

Although much ado is made about the application of CE marking to a machine, the mechanics of the process are in reality fairly straightforward, each phase being broken down into eight steps (as outlined in Table 13.4).

Fig. 13.1 The CE mark

Table 13.4 The steps to CE marking

Step 1	*Decide* if the product is 'a machine', as defined in The Machinery Directive document itself.
Step 2	*Check* whether the machine appears on the list of those that are excluded from the requirements of The Machinery Directive (see Table 13.1), or whether it falls under the jurisdiction of any other European directive (such as, for example, the Low Voltage or Medical Devices Directive), in which case The Machinery Directive would not apply.
Step 3	*Perform* a risk assessment to make sure that all potential safety risks have been eliminated or minimized during the design and construction of the machine.
Step 4	*Demonstrate and record* that the Essential Safety Requirements (ESRs) in Annex I of The Machinery Directive (see Table 13.3) have been complied with.
	OR
	Demonstrate and record that the machine has been designed and manufactured to a European Harmonized Standard, in which case it will carry a presumption of conformity with The Machinery Directive (including the ESRs).
Step 5	*Check* if the machine is specifically listed in Annex IV of The Machinery Directive. If it is, then it needs special test procedures applied to it and independent certification by a Notified Body.
Step 6	*Assemble* a technical file for the machine. This must comply with the required contents as set out in Annex V of The Machinery Directive.
Step 7	*Draw up and sign* a declaration of conformity in accordance with Annex II of The Machinery Directive.
Step 8	*Affix the CE marking* to the machine, on its nameplate. This needs to be done in accordance with Annex III of The Machinery Directive which specifies the minimum height of letters (>5 mm) and the precise way in which the mark must be displayed.

13.5 The technical file

This is a file of information compiled by the manufacturer. The principle of The Machinery Directive is that the contents of the technical file must meet five specific requirements (see Table 13.5) and, by implication, provide adequate coverage of information relevant to health and safety aspects of the machinery. Table 13.6 shows the typical content of a technical file.

Table 13.5 Five requirements of the technical file

1. The data in the technical file must be *relevant*.
2. The information must be *complete and correct*.
3. The information must be *available on time*. (The Machinery Directive does not require that the technical file be physically present at all times, but rather that the information must be made 'available within a period of time commensurate with its importance'.)
4. All information on the machinery must be *consistent*. Information in instructions, advertising materials, and the technical file may not conflict.
5. The information must be *retrievable*.

Table 13.6 Typical content of a technical file

- Manufacturer's name and address
- Machine identification and description
- General arrangement and/or assembly drawing
- Detailed engineering drawings
- Detailed technical calculations
- Detailed test results
- Relevant technical specifications
- A list of relevant European Harmonized Standards and/or reports demonstrating compliance with the essential safety requirements (ESRs)
- Detailed operating instructions for the machine
- Quality assurance procedures
- Information on methods used to eliminate hazards
- Information on methods of risk assessment, i.e. measures/methods used and their results and conclusions
- Details of agreements with any third parties relevant to the design, manufacture, and testing of the machine
- Relevant commercial (e.g. advertising) documentation

The actual level of detail in the technical file depends on the individual nature of the machine. It has to be kept for ten years after the last product has been produced.

13.6 The declaration of conformity

Annex II of The Machinery Directive specifies that machines must be supplied with a 'declaration of conformity'. This is basically a certificate provided by the machine's manufacturer (or, in some cases, importer), stating that all the requirements of The Machinery Directive have been met. Table 13.7 shows the minimum acceptable content of the declaration of conformity and Table 13.8 gives a typical pro-forma example. There is special provision where a manufactured component is supplied to be incorporated into an assembly or large machine manufactured by someone else. In this case a 'certificate of incorporation' replaces the declaration of conformity; see Annex II of The Machinery Directive for details.

Table 13.7 Content of the EC declaration of conformity

An EC declaration of conformity must:

(a) state the business name and full address of:
  - i) the responsible person; and
  - ii) where that person is not the manufacturer, of the manufacturer;

(b) contain a description of the machinery to which the declaration relates which, without prejudice to the generality of the foregoing, includes, in particular:
  - i) its make;
  - ii) type; and
  - iii) serial number;

(c) indicate all relevant provisions with which the machinery complies;

(d) state, in the case of relevant machinery in relation to which an EC type examination certificate has been issued, the name and address of the approved body that issued the certificate and the number of such certificate;

(e) state, in the case of relevant machinery in respect of which a technical file has been drawn up, the name and address of the approved body to which the file has been sent or which has drawn up a certificate of adequacy for the file, as the case may be;

(f) specify (as appropriate) the transposed harmonized standards used;

(g) specify (as appropriate) the national standards and any technical specifications used; and

(h) identify the person authorized to sign the declaration on behalf of the responsible person.

Table 13.8 Declaration of conformity (typical)

Declaration of Conformity

We, *(manufacturer)*

of *(address)*

declare that the machinery

Make:

Type:

Model:

Serial Number:

Year of Construction:

has been manufactured using the following transposed Harmonised European Standards and technical specifications:

and is in conformity with: *(directives to which the product conforms)*

e.g. The Machinery Directive 89/392/EEC as amended by Directive 91/368/EEC, Directive 93/44/EEC, Directive 93/68/EEC, and Directive 98/37/EC,

the Low Voltage Directive 73/23/EEC as amended by Directive 93/68/EEC,

and the EMC Directive 89/336/EEC as amended by Directive 92/31/EEC and 93/68/EEC.

Signed in: *(place)*

on the: *(date)*

Signature:

Name: *(responsible person)*

Position:

Table 13.9 Typical layout of machinery instructions

1. PRODUCT INFORMATION

- Supplier information:
  - the product (type, mark);
  - the supplier;
  - the address.
- Accurately describe the intended use of the product as well as the environment and circumstances under which it should be used. Draw attention to safety aspects of the usage of the machinery (for example: 'only to be used by authorized users').
- In a separate chapter headed 'Safety', describe the safety measures applied:
  - applicable guidelines;
  - explanation of safety symbols, pictograms, warnings, etc. used;
  - possible dangers when:
    - safety instructions are not followed;
    - machinery is not used by trained personnel;
    - machinery is modified, adapted or changed;
  - dangerous user conditions (for example weather conditions);
  - what to do if …;
  - state when guarantee ends.
- Technical specifications:
  - noise levels;
  - vibration levels;
  - heat;
  - radiation.
- Customer service:
  - ordering parts;
  - where customers can report complaints, requirements, deficiencies.

2. INSTALLATION

- Transport instructions;
- Assembly instructions:
  - requirements for the machinery foundation, use of shock absorbers;
  - lifting accessories;
  - personnel (for example: required knowledge, education level).
- Connection requirements (legal requirements, standards);

Table 13.9 Cont.

- Connection of:
  - electrical system;
  - pneumatic system;
  - hydraulic system.

3. USAGE

- Putting into service (first use):
  - preparation (filling oil reservoirs, switching on electrical circuits, etc.);
  - user instructions for the operator;
  - personal protection equipment (safety glasses, helmet, etc.);
  - indicate potential dangers to onlookers.
- The authorized user:
  - type of person (required ability, experience, knowledge, etc.)
  - what additional training, instructions.
- Machinery operation:
  - description of the control panel (and/or software interface);
  - tools/accessories;
  - running in production:
    - taking working sequence into account;
    - possible fault and warning signals.
- Stopping:
  - taking sequence into account;
  - safety steps to be taken
    - waste oil;
    - pressurized air, hydraulics;
    - cleaning;
  - emergency stop.
- Trouble-shooting:
  - to be carried out by operator;
  - to be carried out by service personnel.

4. MAINTENANCE

- Who may maintain the machinery (and what can or cannot be done ...):
  - operator;
  - user's trained personnel (knowledge levels, training);
  - supplier's trained personnel.
- Tests and checks.
- Dangers during maintenance/testing.

Table 13.9 Cont.

- Special provisions:
  - switching on/off;
  - control during operation.
- Safety measures.
- The use of special tools supplied for safe maintenance, cleaning, etc. of the machinery.

5. ACCESSORIES

- Tools, equipment, accessories.
- Safety components (lifting bolts/eyes, etc.):
  - Machinery Directive requirements.
- Original 'spare parts' (including order information).
- Additional tools, etc. available.
- Draw attention to:
  - expertise of personnel;
  - safety measures.
- Timing (intervals, duration, etc.).

Machinery instructions

Under The Directive, machinery has to be supplied with adequate instructions for its use. Table 13.9 shows typical content that is generally required.

13.7 The role of technical standards

Traditionally, most EU countries had (and in most cases still have) their own well-established product standards for all manner of manufactured products, including many types of rotating (and other) machinery. Inevitably, these standards differ in their technical and administrative requirements, and often in the fundamental way that compliance of products with the standards is assured.

Harmonized standards

Harmonized standards are European standards produced (in consultation with member states) by the European standards organizations CEN/CENELEC. There is a Directive 98/34/EC that explains the formal status of these harmonized standards. Harmonized standards have to be 'transposed' by each EU country, which means that they must be made available as national standards and that any conflicting standards have to be withdrawn within a given time period.

A key point about harmonized standards is that any product that complies with the standards is automatically assumed to conform to the Essential Safety Requirements of the New Approach European directive relevant to the particular product. This is known as the 'presumption of conformity'. Once a national standard is transposed from a harmonized standard, then the presumption of conformity is carried with it.

Note that the following terms appear in various directives, guidance notes, etc.: (They are all exactly the same thing.)

- essential safety requirements (ESRs);
- essential requirements;
- essential health and safety requirements (EHSRs).

Compliance with harmonized standards is not compulsory, it is voluntary. Compliance does, however, infer that a product meets the essential safety requirements (ESRs) of a relevant directive and the product can then carry the CE mark. Table 13.10 shows the index of ESRs.

Table 13.10 Index of ESRs

The Machinery Directive contains a detailed list of Essential Safety Requirements (ESRs) (actually termed the Essential Health and Safety Requirements) relating to the design and construction of machinery and safety components. The full text (Annex 1 of The Directive) covers more than 30 pages.

This table shows an index of the ESRs, along with the reference number under which they are listed in Annex 1.

1 ESSENTIAL HEALTH AND SAFETY REQUIREMENTS

1.1 General remarks

1.1.0 Definitions

1.1.1 Principles of safety integration

1.1.2 Materials and products

1.1.3 Lighting

1.1.4 Design of machinery to facilitate its handling

1.2 Controls

1.2.0 Safety and reliability of control systems

1.2.1 Control devices

1.2.2 Starting

1.2.3 Stopping device

1.2.4 Mode selection

1.2.5 Failure of the power supply

Table 13.10 Cont.

1.2.6	Failure of the control circuit
1.2.7	Software
1.3 Protection against mechanical hazards	
1.3.0	Stability
1.3.1	Risk of break-up during operation
1.3.2	Risks due to falling or ejected objects
1.3.3	Risks due to surfaces, edges, or angles
1.3.4	Risks related to combined machinery
1.3.5	Risks relating to variations in the rotational speed of tools
1.3.6	Prevention of risks related to moving parts
1.3.7	Choice of protection against risks related to moving parts
1.4 Required characteristics of guards and protection devices	
1.4.0	General requirement
1.4.1	Special requirements for guards
1.4.2	Special requirements for protection devices
1.5 Protection against other hazards	
1.5.0	Electricity supply
1.5.1	Static electricity
1.5.2	Energy supply other than electricity
1.5.3	Errors of fitting
1.5.4	Extreme temperatures
1.5.5	Fire
1.5.6	Explosion
1.5.7	Noise
1.5.8	Vibration
1.5.9	Radiation
1.5.10	External radiation
1.5.11	Laser equipment
1.5.12	Emissions of dust, gases, etc.
1.5.13	Risk of being trapped in a machine
1.5.14	Risk of slipping, tripping, or falling
1.6 Maintenance	
1.6.0	Machinery maintenance
1.6.1	Machine operating position and servicing points
1.6.2	Isolation of energy sources
1.6.3	Operator intervention

Table 13.10 Cont.

1.6.4	Cleaning of internal parts
1.7 Indicators	
1.7.0	Information devices
1.7.1	Warning devices
1.7.2	Warning of residual risks
1.7.3	Marking
1.7.4	Instructions
2 ESSENTIAL HEALTH AND SAFETY REQUIREMENTS FOR CERTAIN CATEGORIES OF MACHINERY	
2.1 Agri-foodstuffs machinery	
2.2 Portable hand-held and/or hand-guided machinery	
2.3 Machinery for working wood and analogous materials	
3 ESSENTIAL HEALTH AND SAFETY REQUIREMENTS TO OFFSET THE PARTICULAR HAZARDS DUE TO THE MOBILITY OF MACHINERY	
3.1 General	
3.1.1	Definition
3.1.2	Lighting
3.1.3	Design of machinery to facilitate its handling
3.2 Work stations	
3.2.1	Driving position
3.2.2	Seating
3.2.3	Other places
3.3 Controls	
3.3.1	Control devices
3.3.2	Starting/moving
3.3.3	Travelling function
3.3.4	Movement of pedestrian-controlled machinery
3.3.5	Control circuit failure
3.4 Protection against mechanical hazards	
3.4.1	Uncontrolled movements
3.4.2	Risk of break-up during operation
3.4.3	Rollover
3.4.4	Falling objects
3.4.5	Means of access
3.4.6	Towing devices
3.4.7	Transmission of power between self-propelled machinery (or tractor) and recipient machinery

Table 13.10 Cont.

3.4.8 Moving transmission parts

3.5 Protection against other hazards

3.5.1 Batteries

3.5.2 Fire

3.5.3 Emissions of dust, gases, etc.

3.6 Indications

3.6.1 Signs and warnings

3.6.2 Marking

3.6.3 Instruction handbook

4 ESSENTIAL HEALTH AND SAFETY REQUIREMENTS TO OFFSET THE PARTICULAR HAZARDS DUE TO A LIFTING OPERATION

4.1 General remarks

4.1.1 Definitions

4.1.2 Protection against mechanical hazards

4.1.2.1 Risks due to lack of stability

4.1.2.2 Guide rails and rail tracks

4.1.2.3 Mechanical strength

4.1.2.4 Pulleys, drums, chains, or ropes

4.1.2.5 Separate lifting accessories

4.1.2.6 Control of movements

4.1.2.7 Handling of loads

4.1.2.8 Lighting

4.2 Special requirements for machinery whose power source is other than manual effort

4.2.1 Controls

4.2.1.1 Driving position

4.2.1.2 Seating

4.2.1.3 Control devices

4.2.1.4 Loading control

4.2.2 Installation guided by cables

4.2.3 Risks to exposed persons. Means of access to driving position and intervention points

4.2.4 Fitness for purpose

4.3 Marking

4.3.1 Chains and ropes

4.3.2 Lifting accessories

4.3.3 Machinery

Table 13.10 Cont.

4.4 Instruction handbook
- 4.4.1 Lifting accessories
- 4.4.2 Machinery

5 *ESSENTIAL HEALTH AND SAFETY REQUIREMENTS FOR MACHINERY INTENDED FOR UNDERGROUND WORK*

5.1 Risks due to lack of stability

5.2 Movement

5.3 Lighting

5.4 Control devices

5.5 Stopping

5.6 Fire

5.7 Emissions of dust, gases, etc.

6 *ESSENTIAL HEALTH AND SAFETY REQUIREMENTS TO OFFSET THE PARTICULAR HAZARDS DUE TO THE LIFTING OR MOVING OF PERSONS*

6.1 General
- 6.1.1 Definition
- 6.1.2 Mechanical strength
- 6.1.3 Loading control for types of device moved by power other than human strength

6.2 Controls
- 6.2.1 Where safety requirements do not impose other solutions

6.3 Risks of persons falling from the carrier

6.4 Risks of the carrier falling or overturning

6.5 Markings

National standards

EU countries are at liberty to keep their national product standards if they wish. Products manufactured to these do not, however, carry the presumption of conformity with relevant directives, hence the onus is on the manufacturer to prove compliance with the ESRs to a Notified Body (on a case-by-case basis). Once compliance has been demonstrated, then the product can carry the CE mark.

The process of harmonization of standards is ongoing. Inevitably, given the high number and complexity of national standards that exist, harmonized standards are being produced more quickly in some technical fields than in others. Table 13.11 lists the current ones for common types of rotating equipment and Table 4.4 some covering wider, generic technical ones such as vibration, noise, etc.

Table 13.11 Some harmonized standards relevant to The Machinery Directive

Organization	*Reference*	*Title of the harmonized standards*
CEN	EN 115/A1: 1998	Safety rules for the construction and installation of escalators and passenger conveyors.
CEN	EN 201/A1: 2000	Rubber and plastics machines – Injection moulding machines – Safety requirements.
CEN	EN 289: 1993	Rubber and plastics machinery – Compression and transfer moulding presses – Safety requirements for the design.
CEN	EN 292-1: 1991	Safety of machinery – Basic concepts, general principles for design – Part 1: Basic terminology, methodology.
CEN	EN 292-2/A1: 1995	Safety of machinery – Basic concepts, general principles for design – Part 2: Technical principles and specifications.
CEN	EN 294: 1992	Safety of machinery – Safety distance to prevent danger zones being reached by the upper limbs.
CEN	EN 349: 1993	Safety of machinery – Minimum gaps to avoid crushing of parts of the human body.
CEN	EN 418: 1992	Safety of machinery – Emergency stop equipment, functional aspects – Principles for design.
CEN	EN 457: 1992	Safety of machinery – Auditory danger signals – General requirements, design, and testing (ISO 7731: 1986, modified).
CEN	EN 474-1/A1: 1998	Earth-moving machinery – Safety – Part 1: General requirements.
CEN	EN 547-1: 1996	Safety of machinery – Human body measurements – Part 1: Principles for determining the dimensions required for openings for the whole body access into machinery.

Table 13.11 Cont.

CEN	EN 547-2: 1996	Safety of machinery – Human body measurements – Part 2: Principles for determining the dimensions required for access openings.
CEN	EN 547-3: 1996	Safety of machinery – Human body measurements – Part 3: Anthropometric data.
CEN	EN 563: 1994	Safety of machinery – Temperatures of touchable surfaces – Ergonomics data to establish temperature limit values for hot surfaces.
CEN	EN 574: 1996	Safety of machinery – Two-hand control devices – Functional aspects – Principles for design.
CEN	EN 614-1: 1995	Safety of machinery – Ergonomic design principles – Part 1: Terminology and general principles.
CEN	EN 626-1: 1994	Safety of machinery – Reduction of risks to health from hazardous substances emitted by machinery – Part 1: Principles and specifications for machinery manufacturers.
CEN	EN 626-2: 1996	Safety of machinery – Reduction of risk to health from hazardous substances emitted by machinery – Part 2: Methodology leading to verification procedures.
CEN	EN 692: 1996	Mechanical presses – Safety.
CEN	EN 809: 1998	Pumps and pump units for liquids – Common safety requirements.
CEN	EN 842: 1996	Safety of machinery – Visual danger signals – General requirements, design, and testing.
CEN	EN 894-1: 1997	Safety of machinery – Ergonomics requirements for the design of displays and control actuators – Part 1: General principles for human interactions with displays and control actuators.
CEN	EN 894-2: 1997	Safety of machinery – Ergonomics requirements for the design of displays and control actuators – Part 2: Displays.

Table 13.11 Cont.

CEN	EN 953: 1997	Safety of machinery – Guards – General requirements for the design and construction of fixed and movable guards.
CEN	EN 954-1: 1996	Safety of machinery – Safety-related parts of control systems – Part 1: General principles for design.
CEN	EN 981: 1996	Safety of machinery – System of auditory and visual danger and information signals.
CEN	EN 982: 1996	Safety of machinery – Safety requirements for fluid power systems and their components – Hydraulics.
CEN	EN 983: 1996	Safety of machinery – Safety requirements for fluid power systems and their components – Pneumatics.
CEN	EN 999: 1998	Safety of machinery – The positioning of protective equipment in respect of approach speeds of parts of the human body.
CEN	EN 1032/A1: 1998	Mechanical vibration – Testing of mobile machinery in order to determine the whole-body vibration emission value – General – Amendment 1.
CEN	EN 1037: 1995	Safety of machinery – Prevention of unexpected start-up.
CEN	EN 1050: 1996	Safety of machinery – Principles for risk assessment.
CEN	EN 1088: 1995	Safety of machinery – Interlocking devices associated with guards – Principles for design and selection.
CEN	EN 1299: 1997	Mechanical vibration and shock – Vibration isolation of machines – Information for the application of source isolation.
CEN	EN 1760-1: 1997	Safety of machinery – Pressure-sensitive protective devices – Part 1: General principles for the design and testing of pressure-sensitive mats and pressure-sensitive floors.

Table 13.11 Cont.

CEN	EN ISO 3743-1: 1995	Acoustics – Determination of sound power levels of noise sources – Engineering methods for small, movable sources in reverberant fields – Part 1: Comparison method for hard-walled test rooms (ISO 3743-1: 1994).
CEN	EN ISO 3743-2: 1996	Acoustics – Determination of sound power levels of noise sources using sound pressure – Engineering methods for small, movable sources in reverberant fields – Part 2: Methods for special reverberation test rooms (ISO 3743-2: 1994).
CEN	EN ISO 3744: 1995	Acoustics – Determination of sound power levels of noise sources using sound pressure – Engineering method in an essentially free field over a reflecting plane (ISO 3744: 1994).
CEN	EN ISO 3746: 1995	Acoustics – Determination of sound power levels of noise sources using sound pressure – Survey method using an enveloping measurement surface over a reflecting plane (ISO 3746: 1995).
CEN	EN ISO 4871: 1996	Acoustics – Declaration and verification of noise emission values of machinery and equipment (ISO 4871: 1996).
CEN	EN ISO 9614-1: 1995	Acoustics – Determination of sound power levels of noise sources using sound intensity – Part 1: Measurement at discrete points (ISO 9614-1: 1993).
CEN	EN ISO 11200: 1995	Acoustics – Noise emitted by machinery and equipment – Guidelines for the use of basic standards for the determination of emission sound pressure levels at a work station and at other specified positions (ISO 11200: 1995).

Table 13.11 Cont.

CEN	EN ISO 11201: 1995	Acoustics – Noise emitted by machinery and equipment – Measurement of emission sound pressure levels at a work station and at other specified positions – Engineering method in an essentially free field over a reflecting plane (ISO 11201: 1995).
CEN	EN ISO 11202: 1995	Acoustics – Noise emitted by machinery and equipment – Measurement of emission sound pressure levels at a work station and at other specified positions – Survey method in situ (ISO 11202: 1995).
CEN	EN ISO 11203: 1995	Acoustics – Noise emitted by machinery and equipment – Determination of emission sound pressure levels at a work station and at other specified positions from the sound power level (ISO 11203: 1995).
CEN	EN ISO 11204: 1995	Acoustics – Noise emitted by machinery and equipment – Measurement of emission sound pressure levels at a work station and at other specified positions – Method requiring environmental corrections (ISO 11204: 1995).
CEN	EN ISO 11546-1: 1995	Acoustics – Determination of sound insulation performances of enclosures – Part 1: Measurements under laboratory conditions (for declaration purposes) (ISO 11546-1: 1995).
CEN	EN ISO 11546-2: 1995	Acoustics – Determination of sound insulation performances of enclosures – Part 2: Measurements in situ for acceptance and verification purposes (ISO 11546-2: 1995).
CEN	EN ISO 13753: 1998	Mechanical vibration and shock – Hand–arm vibration – Method for measuring the vibration transmissibility of resilient materials when loaded by the hand–arm system (ISO 13753: 1998).

Table 13.11 Cont.

CENELEC	EN 61310-1: 1995	Safety of machinery – Indication, marking, and actuation – Part 1: Requirements for visual, auditory, and tactile signals (IEC 61310-1: 1995).
CENELEC	EN 61310-2: 1995	Safety of machinery – Indication, marking, and actuation – Part 2: Requirements for marking (IEC 61310-2: 1995).
CENELEC	EN 61310-3: 1999	Safety of machinery – Indication, marking, and actuation – Part 3: Requirements for the location and operation of actuators (IEC 61310-3: 1999).
CENELEC	EN 61496-1: 1997	Safety of machinery – Electro-sensitive protective equipment – Part 1: General requirements and tests (IEC 61496-1: 1997).

13.8 The proposed 'amending' directive 95/16/EC

There are currently proposals for an amending directive 95/16/EC, which would effectively 're-cast' the content of the consolidating Machinery Directive 98/37/EC. The stated purpose is to clarify the content of The Machinery Directive and leave it less open to incorrect interpretation. The amendments are based on a set of twelve proposals set out at the beginning of the 95/16/EC document. Table 13.12 outlines some of the changes (refer to the document itself for full details).

Table 13.12 Some changes to The Machinery Directive proposed by 95/16/EC

Articles

Article 1 has been very substantially amended to take account of comments to the effect that not all of the products referred to in The Directive are machines in the strict sense of the word. The new definition takes account of this aspect and clearly identifies partly completed machinery, to which The Directive does not apply in its entirety.

A number of definitions have been added to make it easier to interpret the text. In the case of safety components, it was decided to present an exhaustive list of machinery rather than a definition (the text of Directive 98/37/EC contains such a definition and it has given rise to many problems of interpretation). In order to take account of technological development, the Machinery Committee set up by The Directive will have the powers to amend this list.

Annex I – Essential health and safety requirements

The essential health and safety requirements set out in Annex I have not been fundamentally changed; the numbering of the various points has been maintained wherever possible. Many of the changes to the original text concern the drafting.

Annex II – Declarations

The contents of the declarations described in Annex II have been amended to take account of the incorporation of safety components into machinery. There are now only two types of declaration: the 'EC conformity declaration' for all machinery and the 'declaration of incorporation' for partly completed machinery.

Annex IV – Categories of potentially hazardous machinery

The list in Annex IV of machinery regarded as most hazardous has been amended to take account of the difficulties of interpreting the existing list.

Annexes V and VIII on partly completed machinery and intrinsically safe machinery

A specific annex setting out the assembly instructions for partly completed machinery has been added (Annex V). The same goes for the conformity assessment of a machine not exhibiting any intrinsic health and safety hazard (Annex VIII).

Annexes VI, VII, IX, and XI on conformity assessment

The content of these annexes, which corresponds to the modules set out in Decision 93/465/EEC (Annexes VI, VII, X, and XI), has been maintained, although the wording has been changed to make them easier to use.

Table 13.12 Cont.

The technical file which is included in a number of modules is now the subject of a separate annex (Annex VI).

Annex IX on the adequacy of a machine in respect of harmonized standards has been added to take account of the practice in the 1989 Directive, which was drafted before the modules were adopted. This procedure is a major simplification for manufacturers who have elected to manufacture their machinery in accordance with harmonized standards.

In Annex IX (adequacy in respect of harmonized standards) and Annex X (EC type-examination), it has been specified that the Notified Body must keep its technical file for 15 years. This detail does not appear in the modules.

Annex XI on fully quality assurance has been amended, relative to the corresponding module, to make it clear that the manufacturer must, for each of the machines he manufactures, possess a technical file so as to be able to respond to any reasoned request from a member state which might consider that the machinery in question is defective.

13.9 Useful references and standards

New Approach directives and harmonized standards are listed on:
http://europa.eu.int/comm./enterprise/newapproach/standardization/harmstds/

Information can be obtained from:
Standardization unit – European Commission
Mr D Herbert
European Commission
rue de la Loi 200
B-1049
Brussels

Contact person: Ingrid.gillisjans@cec.eu.ir
European standards organization

- CEN: infodesk@cenorm.be
- CENELEC: general@cenelec.be
- ETSI: infocentre@etsi.fr

Machinery Directive 98/37/EC

The general website, giving access to all the text (including Annexes I–VII), is:
http://europa.eu.int/comm./enterprise/mechan_equipment/machinery/guide/content.htm

Information on The Directive can be obtained from:

- EC-DG ENTR G3
- Mr Van Gheluwe, Tel: 00 32 (2) 296 09 64, Fax: 00 32 (2) 296 62 73
- e-mail: machinery@cec.eu.int

CHAPTER 14

Organizations and Associations

The following table shows some major European and American associations and organizations relevant to rotating equipment activities.

Acronym	*Organization*	*Contact*
ABMA	American Bearing Manufacturers Association 2025 M Street NW Suite 800 Washington DC 20036	Tel: 00 1 (202) 367 1155 Fax: 00 1 (202) 367 2155 www.abma-dc.org
ABS	American Bureau of shipping (UK) ABS House 1 Frying Pan Alley London E1 7HR	Tel: +44 (0)20 7247 3255 Fax: +44 (0)20 7377 2453 www.eagle.org
ACEC	American Consulting Engineers Council 1015, 15th St NW #802 Washington DC 20005	Tel: 00 1 (202) 347 7474 Fax: 00 1 (202) 898 0068 www.acec.org
AEAT	AEA Technology plc (UK) Harwell Didcot Oxon OX11 0QT	Tel: +44 (0)1235 821111 Fax: +44 (0)1235 432916 www.aeat.co.uk
AGMA	American Gear Manufacturers Association 1500 King St Suite 201 Alexandria VA 22314	Tel: 00 1 (703) 684 0211 Fax: 00 1 (703) 684 0242 www.agma.org

ANS	American Nuclear Society 555 N. Kensington Ave La Grange Park IL 60526	Tel: 00 1 (708) 352 6611 Fax: 00 1 (708) 352 0499 www.ans.org
ANSI	American National Standards Institute 11, W. 42nd St New York NY 10036	Tel: 00 1 (212) 642 4900 Fax: 00 1 (212) 398 0023 www.ansi.org
API	American Petroleum Institute 1220 L St NW Washington DC 20005	Tel: 00 1 (202) 682 8000 Fax: 00 1 (202) 682 8232 www.api.org
ASERCOM	Association of European Refrigeration Compressor Manufacturers C/O Copeland GmbH Eichborndamm 141-175 D-1000 Berlin Germany	Tel: 00 (49) 30 419 6352 Fax: 00 (49) 30 419 6205 www.hvacmall.com
ASHRAE	American Society of Heating, Refrigeration and Air Conditioning Engineers 1791 Tullie Circle NE Atlanta GA 30329	Tel: 00 1 (404) 636 8400 Fax: 00 1 (404) 321 5478 www.ashrae.org
ASME	American Society of Mechanical Engineers 3, Park Ave New York NY 10016-5990	Tel: 00 1 (973) 882 1167 Fax: 00 1 (973) 882 1717 www.asme.org
ASNT	American Society for Non-Destructive Testing 1711 Arlington Lane Columbus OH 43228-0518	Tel: 00 1 (614) 274 6003 Fax: 00 1 (614) 274 6899 www.asnt.org
ASTM	American Society for Testing of Materials 100, Barr Harbor Drive W Conshohocken PA 19428-2959	Tel: 00 1 (610) 832 9585 Fax: 00 1 (610) 832 9555 www.ansi.org

AWS	American Welding Society 550 NW Le Jeune Rd Miami FL 33126	Tel: 00 1 (305) 443 9353 Fax: 00 1 (305) 443 7559 www.awweld.org
AWWA	American Water Works Association Inc 6666 W Quincy Ave Denver CO 80235	Tel: 00 1 (303) 794 7711 Fax: 00 1 (303) 794 3951 www.awwa.org
BCAS	British Compressed Air Society 33-34 Devonshire St London W1G 6YP	Tel: +44 (0)20 7935 2464 Fax: +44 (0)20 7935 2464 www.britishcompressedair. co.uk
BCEMA	British Combustion Equipment Manufacturers Association The Fernery Market Place Midhurst W Sussex GU29 9DP	Tel: +44 (0)1730 812782 Fax: +44 (0)1730 813366 www.bcema.co.uk
BFPA	British Fluid Power Association Cheriton House Cromwell Business Park Chipping Norton Oxon OX7 5SR	Tel: +44 (0)1608 647900 Fax: +44 (0)1608 647919 www.bfpa.co.uk
BGA	British Gear Association Suite 43 Inmex Business Park Shobnall Rd Burton on Trent Staffordshire DE14 2AU	Tel: +44 (0)1283 515521 Fax: +44 (0)1283 515841 www.bga.org.uk
BIE	British Inspecting Engineers Chatsworth Technology Park Dunston Road Chesterfield D41 8XA	Tel: +44 (0)1246 260260 Fax: +44 (0)1246 260919 www.bie-international.com

B.Inst.NDT	British Institute of Non Destructive Testing 1 Spencer Parade Northampton NN1 5AA	Tel: +44 (0)1604 259056 Fax: +44 (0)1604 231489 www.bindt.org
BPMA	British Pump Manufacturers Association The McLaren Building 35 Dale End Birmingham B4 7LN	Tel: +44 (0)121 200 1299 Fax: +44 (0)121 200 1306 www.bpma.org.uk
BSI	British Standards Institution Marylands Avenue Hemel Hempstead Herts HP2 4SQ	Tel: +44 (0)1442 230442 Fax: +44 (0)1442 231442 www.bsi.org.uk
BVAMA	British Valve and Actuator Manufacturers Association The McLaren Building 35 Dale End Birmingham B4 7LN	Tel: +44 (0)121 200 1297 Fax: +44 (0)121 200 1308 www.bvama.org.uk
CEN	European Committee for Standardisation (Belgium) 36, rue de Stassart B-1050 Brussels Belgium	Tel: 00 (32) 2 550 08 11 Fax: 00 (32) 2 550 08 19 www.cenorm.be www.newapproach.org
DNV	Det Norske Veritas (UK) Palace House 3 Cathedral Street London SE1 9DE	Tel: +44 (0)20 7357 6080 Fax: +44 (0)20 357 76048 www.dnv.com
DTI	DTI STRD 5 (UK) Peter Rutter STRD5 Department of Trade and Industry, Room 326 151 Buckingham Palace Road London SW1W 9SS	Tel: +44 (0)20 7215 1437 www.dti.gov.uk/strd

DTI	DTI Publications Orderline (UK)	Tel: +44 (0)870 1502 500 Fax: +44 (0)870 1502 333
EC	The Engineering Council (UK) 10 Maltravers Street London WC2R 3ER	Tel: +44 (0)20 7240 7891 Fax: +44 (0)20 7240 7517 www.engc.org.uk
EIS	Engineering Integrity Society (UK) 5 Wentworth Avenue Sheffield S11 9QX	Tel: +44 (0)114 262 1155 Fax: +44 (0)114 262 1120 www.demon.co.uk/e-i-s
EMA	Engine Manufacturers Association (USA) 2 N LaSalle St Suite 2200 Chicago IL 60602	Tel: 00 1 (312) 827 8700 Fax: 00 1 (312) 827 8737 www.engine-manufacturers.org
FCI	Fluid Controls Institute Inc (USA) PO Box 1485 Pompano Beach FL 33061	Tel: 00 1 (216) 241 7333 Fax: 00 1 (216) 241 0105 www.fluidcontrolsinstitute.org
FMG	Factory Mutual Global (USA) Westwood Executive Center 100 Lowder Brook Drive Suite 1100 Westwood MA 02090-1190	Tel: 00 1 (781) 326 5500 Fax: 00 1 (781) 326 6632 www.fmglobal.com
HI	Hydraulic Institute (USA) 9 Sylvian Way Parsippany NJ 07054	Tel: 00 1 (973) 267 9700 Fax: 00 1 (973) 267 9055 www.pumps.org
HMSO	Her Majesty's Stationery Office	www.hmso.gov.uk/legis.htm www.hmso.gov.uk/si www.legislation.hmso.gov.uk
HSE	HSE Books (UK) PO Box 1999 Sudbury Suffolk CO10 6FS	Tel: +44 (0)1787 881165 Fax: +44 (0)1787 313995 www.hse.gov.uk/hsehome

HSE	HSE's InfoLine (fax enquiries) HSE Information Centre Broad Lane Sheffield S3 7HQ	Fax: +44 (0)114 289 2333 www.hse.gov.uk/hsehome
HTRI	Heat Transfer Research Inc (USA) 1500 Research Parkway Suite 100 College Station TX 77845	Tel: 00 1 (409) 260 6200 Fax: 00 1 (409) 260 6249 www.htrinet.com
IGTI	International Gas Turbine Institute (ASME) 5775-B Glenridge Dr. #370 Atlanta GA 30328	Tel: 00 1 (404) 847 0072 Fax: 00 1 (404) 847 0151 www.asme.org/igti
IMechE	The Institution of Mechanical Engineers (UK) 1 Birdcage Walk London SW1H 9JJ	Tel: +44 (0)20 7222 7899 Fax: +44 (0)20 7222 4557 www.imeche.org.uk
IoC	The Institute of Corrosion (UK) 4 Leck Street Leighton Buzzard Bedfordshire LU7 9TQ	Tel: +44 (0)1525 851771 Fax: +44 (0)1525 376690 www.icorr.demon.co.uk
IoE	The Institute of Energy (UK) 18 Devonshire Street London W1N 2AU	Tel: +44 (0)20 7580 7124 Fax: +44 (0)20 7580 4420 www.instenergy.org.uk
IoM	The Institute of Materials (UK) 1 Carlton House Terrace London SW1Y 5DB	Tel: +44 (0)20 7451 7300 Fax: +44 (0)20 7839 1702 www.instmat.co.uk
IPLantE	The Institution of Plant Engineers (UK) 77 Great Peter Street Westminster London SW1P 2EZ	Tel: +44 (0)20 7233 2855 Fax: +44 (0)20 7233 2604 www.iplante.org.uk

IQA	The Institute of Quality Assurance (UK) 12 Grosvenor Crescent London SW1X 7EE	Tel: +44 (0)20 7245 6722 Fax: +44 (0)20 7245 6755 www.iqa.org
ISO	International Standards Organization (Switzerland) PO Box 56 CH-1211 Geneva Switzerland	Tel: 00 (22) 749 011 Fax: 00 (22) 733 3430 www.iso.ch
LR	Lloyd's Register (UK) 71 Fenchurch St London EC3M 4BS	Tel: +44 (0)20 7709 9166 Fax: +44 (0)20 7488 4796 www.lrqa.com
MSS	Manufacturers Standardization Society of the Valve and Fittings Industry (USA) 127 Park Street NE Vienna VA 22180-4602	Tel: 00 1 (703) 281 6613 Fax: 00 1 (703) 281 6671 www.mss-hq.com
NACE	National Association of Corrosion Engineers (USA) 1440 South Creek Drive Houston TX 77084-4906	Tel: 00 1 (281) 228 6200 Fax: 00 1 (281) 228 6300 www.nace.org
NFP	National Fire Protection Association (USA) 1, Batterymarch Park PO Box 9101 Quincy MA 02269-9101	Tel: 00 1 (617) 770 3000 Fax: 00 1 (617) 770 0700 www.nfpa.org
NFPA	National Fluid Power Association (USA) 3333 N Mayfair Rd Milwaukee WI 53222-3219	Tel: 00 1 (414) 778 3344 Fax: 00 1 (414) 778 3361 www.nfpa.com
NIST	National Institute of Standards and Technology (USA) 100 Bureau Drive Gaithersburg MD 20899-0001	Tel: 00 1 (301) 975 8205 Fax: 00 1 (301) 926 1630 www.nist.gov

PDA	Pump Distributors Association 5 Chapelfield Orford Woodbridge IP12 2HW	Tel: +44 (0)1394 450181 Fax: +44 (0)1394 450181 www.pda-uk.com
SAE	Society of Automotive Engineers 400 Commonwealth Drive Warrendale PA 10509-6001	Tel: 00 1 (724) 776 4841 Fax: 00 1 (724) 776 5760 www.sae.org
SAFeD	Safety Assessment Federation (UK) Nutmeg House 60 Gainsford Street Butlers Wharf London SE1 2NY	Tel: +44 (0)20 7403 0987 Fax: +44 (0)20 7403 0137 www.safed.co.uk
TUV	TUV (UK) Ltd Surrey House Surrey St Croydon CR9 1XZ	Tel: +44 (0)20 8680 7711 Fax: +44 (0)20 8680 4035 www.tuv-uk.com
UKAS	The United Kingdom Accreditation Service 21-47 High Street Feltham Middlesex TW13 4UN	Tel: +44 (0)20 8917 8554 Fax: +44 (0)20 8917 8500 www.ukas.com
VGT	Verenigning Gas Turbine (Dutch Gas Turbine Association) Burgemeester Verderiaan 13 3544 AD Utrecht PO Box 261 3454 ZM De Meern Netherlands	Tel: 00 (31) 30 669 1966 Fax: 00 (31) 30 669 1969 www.vgt.org/vgt

Index

Related Titles

Other titles by the author

Engineers' Guide to Pressure Equipment	1 86058 298 2
IMechE Engineers' Data Book – Second Edition	1 86058 248 6
A Guide to Presenting Technical Information	1 86058 249 4
A Practical Guide to Engineering Failure Investigation	1 86058 086 6
Handbook of Mechanical Works Inspection	1 86058 047 5

Other titles of interest

Compressors and their Systems	(IMechE Conference)	1 86058 330 X
Guide to European Pumps and Pumping – Second Edition	Edited by B Nesbitt	1 86058 286 9
Vibrations in Rotating Machinery	IMechE Conference	1 86058 273 7
Reliability of Sealing Systems for Rotating Machinery	IMechE Seminar	1 86058 245 1
Process Pump Selection – A Systems Approach	By J Davidson and O von Bertele	1 86058 180 3
Couplings and Shaft Alignment	By M J Neale, P Needham, and R Horrell	1 86058 170 6
Process Fan and Compressor Selection	Edited by J Davidson and O von Bertele	0 85298 825 7

For the full range of titles published, by Professional Engineering Publishing contact:
Sales Department, Professional Engineering Publishing Limited,
Northgate Avenue, Bury St Edmunds, Suffolk IP32 6BW, UK
Tel: +44 (0) 1284 724384; Fax: +44 (0) 1284 718692
E-mail: sales@pepublishing.com www. pepublishing.com

Printed and bound in the UK by
CPI Antony Rowe, Eastbourne

Printed and bound by CPI Group (UK) Ltd, Croydon, CR0 4YY

16/04/2025

14658543-0001